FREE Test Taking Tips DVD Offer

To help us better serve you, we have developed a Test Taking Tips DVD that we would like to give you for FREE. **This DVD covers world-class test taking tips that you can use to be even more successful when you are taking your test.**

All that we ask is that you email us your feedback about your study guide. Please let us know what you thought about it – whether that is good, bad or indifferent.

To get your **FREE Test Taking Tips DVD**, email freedvd@studyguideteam.com with "FREE DVD" in the subject line and the following information in the body of the email:

  a. The title of your study guide.

  b. Your product rating on a scale of 1-5, with 5 being the highest rating.

  c. Your feedback about the study guide. What did you think of it?

  d. Your full name and shipping address to send your free DVD.

If you have any questions or concerns, please don't hesitate to contact us at freedvd@studyguideteam.com.

Thanks again!

TABE Test Study Guide 2020 and 2021
TABE 11/12 Study Guide and Practice Questions for the Test of Adult Basic Education 11 and 12
[3rd Edition]

Test Prep Books

Table of Contents

Quick Overview

As you draw closer to taking your exam, effective preparation becomes more and more important. Thankfully, you have this study guide to help you get ready. Use this guide to help keep your studying on track and refer to it often.

This study guide contains several key sections that will help you be successful on your exam. The guide contains tips for what you should do the night before and the day of the test. Also included are test-taking tips. Knowing the right information is not always enough. Many well-prepared test takers struggle with exams. These tips will help equip you to accurately read, assess, and answer test questions.

A large part of the guide is devoted to showing you what content to expect on the exam and to helping you better understand that content. In this guide are practice test questions so that you can see how well you have grasped the content. Then, answer explanations are provided so that you can understand why you missed certain questions.

Don't try to cram the night before you take your exam. This is not a wise strategy for a few reasons. First, your retention of the information will be low. Your time would be better used by reviewing information you already know rather than trying to learn a lot of new information. Second, you will likely become stressed as you try to gain a large amount of knowledge in a short amount of time. Third, you will be depriving yourself of sleep. So be sure to go to bed at a reasonable time the night before. Being well-rested helps you focus and remain calm.

Be sure to eat a substantial breakfast the morning of the exam. If you are taking the exam in the afternoon, be sure to have a good lunch as well. Being hungry is distracting and can make it difficult to focus. You have hopefully spent lots of time preparing for the exam. Don't let an empty stomach get in the way of success!

When travelling to the testing center, leave earlier than needed. That way, you have a buffer in case you experience any delays. This will help you remain calm and will keep you from missing your appointment time at the testing center.

Be sure to pace yourself during the exam. Don't try to rush through the exam. There is no need to risk performing poorly on the exam just so you can leave the testing center early. Allow yourself to use all of the allotted time if needed.

Remain positive while taking the exam even if you feel like you are performing poorly. Thinking about the content you should have mastered will not help you perform better on the exam.

Once the exam is complete, take some time to relax. Even if you feel that you need to take the exam again, you will be well served by some down time before you begin studying again. It's often easier to convince yourself to study if you know that it will come with a reward!

Test-Taking Strategies

1. Predicting the Answer

When you feel confident in your preparation for a multiple-choice test, try predicting the answer before reading the answer choices. This is especially useful on questions that test objective factual knowledge. By predicting the answer before reading the available choices, you eliminate the possibility that you will be distracted or led astray by an incorrect answer choice. You will feel more confident in your selection if you read the question, predict the answer, and then find your prediction among the answer choices. After using this strategy, be sure to still read all of the answer choices carefully and completely. If you feel unprepared, you should not attempt to predict the answers. This would be a waste of time and an opportunity for your mind to wander in the wrong direction.

2. Reading the Whole Question

Too often, test takers scan a multiple-choice question, recognize a few familiar words, and immediately jump to the answer choices. Test authors are aware of this common impatience, and they will sometimes prey upon it. For instance, a test author might subtly turn the question into a negative, or he or she might redirect the focus of the question right at the end. The only way to avoid falling into these traps is to read the entirety of the question carefully before reading the answer choices.

3. Looking for Wrong Answers

Long and complicated multiple-choice questions can be intimidating. One way to simplify a difficult multiple-choice question is to eliminate all of the answer choices that are clearly wrong. In most sets of answers, there will be at least one selection that can be dismissed right away. If the test is administered on paper, the test taker could draw a line through it to indicate that it may be ignored; otherwise, the test taker will have to perform this operation mentally or on scratch paper. In either case, once the obviously incorrect answers have been eliminated, the remaining choices may be considered. Sometimes identifying the clearly wrong answers will give the test taker some information about the correct answer. For instance, if one of the remaining answer choices is a direct opposite of one of the eliminated answer choices, it may well be the correct answer. The opposite of obviously wrong is obviously right! Of course, this is not always the case. Some answers are obviously incorrect simply because they are irrelevant to the question being asked. Still, identifying and eliminating some incorrect answer choices is a good way to simplify a multiple-choice question.

4. Don't Overanalyze

Anxious test takers often overanalyze questions. When you are nervous, your brain will often run wild, causing you to make associations and discover clues that don't actually exist. If you feel that this may be a problem for you, do whatever you can to slow down during the test. Try taking a deep breath or counting to ten. As you read and consider the question, restrict yourself to the particular words used by the author. Avoid thought tangents about what the author *really* meant, or what he or she was *trying* to say. The only things that matter on a multiple-choice test are the words that are actually in the question. You must avoid reading too much into a multiple-choice question, or supposing that the writer meant something other than what he or she wrote.

5. No Need for Panic

It is wise to learn as many strategies as possible before taking a multiple-choice test, but it is likely that you will come across a few questions for which you simply don't know the answer. In this situation, avoid panicking. Because most multiple-choice tests include dozens of questions, the relative value of a single wrong answer is small. As much as possible, you should compartmentalize each question on a multiple-choice test. In other words, you should not allow your feelings about one question to affect your success on the others. When you find a question that you either don't understand or don't know how to answer, just take a deep breath and do your best. Read the entire question slowly and carefully. Try rephrasing the question a couple of different ways. Then, read all of the answer choices carefully. After eliminating obviously wrong answers, make a selection and move on to the next question.

6. Confusing Answer Choices

When working on a difficult multiple-choice question, there may be a tendency to focus on the answer choices that are the easiest to understand. Many people, whether consciously or not, gravitate to the answer choices that require the least concentration, knowledge, and memory. This is a mistake. When you come across an answer choice that is confusing, you should give it extra attention. A question might be confusing because you do not know the subject matter to which it refers. If this is the case, don't eliminate the answer before you have affirmatively settled on another. When you come across an answer choice of this type, set it aside as you look at the remaining choices. If you can confidently assert that one of the other choices is correct, you can leave the confusing answer aside. Otherwise, you will need to take a moment to try to better understand the confusing answer choice. Rephrasing is one way to tease out the sense of a confusing answer choice.

7. Your First Instinct

Many people struggle with multiple-choice tests because they overthink the questions. If you have studied sufficiently for the test, you should be prepared to trust your first instinct once you have carefully and completely read the question and all of the answer choices. There is a great deal of research suggesting that the mind can come to the correct conclusion very quickly once it has obtained all of the relevant information. At times, it may seem to you as if your intuition is working faster even than your reasoning mind. This may in fact be true. The knowledge you obtain while studying may be retrieved from your subconscious before you have a chance to work out the associations that support it. Verify your instinct by working out the reasons that it should be trusted.

8. Key Words

Many test takers struggle with multiple-choice questions because they have poor reading comprehension skills. Quickly reading and understanding a multiple-choice question requires a mixture of skill and experience. To help with this, try jotting down a few key words and phrases on a piece of scrap paper. Doing this concentrates the process of reading and forces the mind to weigh the relative importance of the question's parts. In selecting words and phrases to write down, the test taker thinks about the question more deeply and carefully. This is especially true for multiple-choice questions that are preceded by a long prompt.

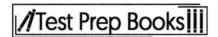

9. Subtle Negatives

One of the oldest tricks in the multiple-choice test writer's book is to subtly reverse the meaning of a question with a word like *not* or *except*. If you are not paying attention to each word in the question, you can easily be led astray by this trick. For instance, a common question format is, "Which of the following is...?" Obviously, if the question instead is, "Which of the following is not...?," then the answer will be quite different. Even worse, the test makers are aware of the potential for this mistake and will include one answer choice that would be correct if the question were not negated or reversed. A test taker who misses the reversal will find what he or she believes to be a correct answer and will be so confident that he or she will fail to reread the question and discover the original error. The only way to avoid this is to practice a wide variety of multiple-choice questions and to pay close attention to each and every word.

10. Reading Every Answer Choice

It may seem obvious, but you should always read every one of the answer choices! Too many test takers fall into the habit of scanning the question and assuming that they understand the question because they recognize a few key words. From there, they pick the first answer choice that answers the question they believe they have read. Test takers who read all of the answer choices might discover that one of the latter answer choices is actually *more* correct. Moreover, reading all of the answer choices can remind you of facts related to the question that can help you arrive at the correct answer. Sometimes, a misstatement or incorrect detail in one of the latter answer choices will trigger your memory of the subject and will enable you to find the right answer. Failing to read all of the answer choices is like not reading all of the items on a restaurant menu: you might miss out on the perfect choice.

11. Spot the Hedges

One of the keys to success on multiple-choice tests is paying close attention to every word. This is never truer than with words like almost, most, some, and sometimes. These words are called "hedges" because they indicate that a statement is not totally true or not true in every place and time. An absolute statement will contain no hedges, but in many subjects, the answers are not always straightforward or absolute. There are always exceptions to the rules in these subjects. For this reason, you should favor those multiple-choice questions that contain hedging language. The presence of qualifying words indicates that the author is taking special care with his or her words, which is certainly important when composing the right answer. After all, there are many ways to be wrong, but there is only one way to be right! For this reason, it is wise to avoid answers that are absolute when taking a multiple-choice test. An absolute answer is one that says things are either all one way or all another. They often include words like *every*, *always*, *best*, and *never*. If you are taking a multiple-choice test in a subject that doesn't lend itself to absolute answers, be on your guard if you see any of these words.

12. Long Answers

In many subject areas, the answers are not simple. As already mentioned, the right answer often requires hedges. Another common feature of the answers to a complex or subjective question are qualifying clauses, which are groups of words that subtly modify the meaning of the sentence. If the question or answer choice describes a rule to which there are exceptions or the subject matter is complicated, ambiguous, or confusing, the correct answer will require many words in order to be expressed clearly and accurately. In essence, you should not be deterred by answer choices that seem excessively long. Oftentimes, the author of the text will not be able to write the correct answer without

offering some qualifications and modifications. Your job is to read the answer choices thoroughly and completely and to select the one that most accurately and precisely answers the question.

13. Restating to Understand

Sometimes, a question on a multiple-choice test is difficult not because of what it asks but because of how it is written. If this is the case, restate the question or answer choice in different words. This process serves a couple of important purposes. First, it forces you to concentrate on the core of the question. In order to rephrase the question accurately, you have to understand it well. Rephrasing the question will concentrate your mind on the key words and ideas. Second, it will present the information to your mind in a fresh way. This process may trigger your memory and render some useful scrap of information picked up while studying.

14. True Statements

Sometimes an answer choice will be true in itself, but it does not answer the question. This is one of the main reasons why it is essential to read the question carefully and completely before proceeding to the answer choices. Too often, test takers skip ahead to the answer choices and look for true statements. Having found one of these, they are content to select it without reference to the question above. Obviously, this provides an easy way for test makers to play tricks. The savvy test taker will always read the entire question before turning to the answer choices. Then, having settled on a correct answer choice, he or she will refer to the original question and ensure that the selected answer is relevant. The mistake of choosing a correct-but-irrelevant answer choice is especially common on questions related to specific pieces of objective knowledge. A prepared test taker will have a wealth of factual knowledge at his or her disposal, and should not be careless in its application.

15. No Patterns

One of the more dangerous ideas that circulates about multiple-choice tests is that the correct answers tend to fall into patterns. These erroneous ideas range from a belief that B and C are the most common right answers, to the idea that an unprepared test-taker should answer "A-B-A-C-A-D-A-B-A." It cannot be emphasized enough that pattern-seeking of this type is exactly the WRONG way to approach a multiple-choice test. To begin with, it is highly unlikely that the test maker will plot the correct answers according to some predetermined pattern. The questions are scrambled and delivered in a random order. Furthermore, even if the test maker was following a pattern in the assignation of correct answers, there is no reason why the test taker would know which pattern he or she was using. Any attempt to discern a pattern in the answer choices is a waste of time and a distraction from the real work of taking the test. A test taker would be much better served by extra preparation before the test than by reliance on a pattern in the answers.

FREE DVD OFFER

Don't forget that doing well on your exam includes both understanding the test content and understanding how to use what you know to do well on the test. We offer a completely FREE Test Taking Tips DVD that covers world class test taking tips that you can use to be even more successful when you are taking your test.

All that we ask is that you email us your feedback about your study guide. To get your **FREE Test Taking Tips DVD**, email freedvd@studyguideteam.com with "FREE DVD" in the subject line and the following information in the body of the email:

- The title of your study guide.
- Your product rating on a scale of 1-5, with 5 being the highest rating.
- Your feedback about the study guide. What did you think of it?
- Your full name and shipping address to send your free DVD.

Introduction to the TABE 11 & 12

Function of the Test

The Tests of Adult Basic Education (TABEs) are a series of tests offered by Data Recognition Corporation as assessment and diagnostic tests for use in the field of adult education. The tests are administered by a wide variety of adult education programs including GED programs, community colleges, adult literacy programs, and correctional institution literacy programs, among others. Most students who take the test do so at the request of the adult education program they are involved with or planning to enter, so that the program can determine each student's capabilities and what sort of curriculum would be most appropriate.

The TABE tests are used by adult education programs across the United States, and often administered by the programs in question to assist with placement and diagnostics within their own program. The typical test taker is an adult with a basic level of education.

This guide is for the TABE 11/12 test. More information about this can be found in the table on the next page.

Test Administration

Generally, test administration is handled by the organization conducting the assessment. Accordingly, the location, date, time, and administrative logistics are determined on a case-by-case basis by those organizations. Likewise, each organization is free to retest students, if and when they see fit.

Test takers with documented disabilities are generally able to receive reasonable accommodations during the TABE testing process. Among other accommodations, a large-print edition, a Braille edition, and an audio edition of the TABE tests are available. Accommodations are arranged through the entity administering the test.

Test Format

TABE tests are offered in a variety of formats, depending on the needs and circumstances of the administering entity. They can be given in a pencil-and-paper format, online via the internet, or on a local computer. Depending on the format used, there are also a variety of test scoring options, including hand scoring, local scanning, and online scoring.

A summary of the content and purpose of some of the most prominent TABE tests follows:

Test	Summary	Purpose
TABE 9 & 10	Battery of subject tests including Reading, Math, and Language	Broad-based adult basic skills assessment
TABE 11 & 12	Is an updated battery of subject tests that align with the new College- and Career-Readiness Standards. It includes Literacy, Reading, Language, and Math	Broad-based adult basic skills assessment
TABE Advanced-Level Tests	Battery of subject tests including Science, Social Studies, Algebra/Geometry, and Writing	Supplemental tests to diagnose readiness for high school equivalency exams
TABE Complete Language Assessment System- English	Examination of English language skills including reading, writing, listening, and speaking	Assessing the language skills of adult English-speakers
TABE Online	An online version of the TABE 9&10 test	Quick evaluation of readiness for education or employment

Scoring

Scoring methodology varies among the different tests offered within the TABE series, but typically, the test taker receives a raw score based on the number of correct answers provided, with no penalty for incorrect answers. The raw score is translated to a scaled score, which represents a certain level of aptitude in adult education. The scaled scores can also be translated to a grade level correlation, which indicates an approximate grade level for the test taker in a given subject so that the student is placed in the appropriate curriculum for their needs.

Recent/Future Developments

Changes have gradually been taking place, including shifting test takers toward computer-based exams and adjusting the test content to meet national college readiness standards. These changes are likely to continue in the future.

Reading

Key Ideas and Details

Cite Strong and Thorough Textual Evidence

<u>Supporting Details</u>
Supporting details help readers better develop and understand the main idea. Supporting details answer questions like *who, what, where, when, why,* and *how*. Different types of supporting details include examples, facts and statistics, anecdotes, and sensory details.

Persuasive and informative texts often use supporting details. In persuasive texts, authors attempt to make readers agree with their point of view, and supporting details are often used as "selling points." If authors make a statement, they should support the statement with evidence in order to adequately persuade readers. Informative texts use supporting details such as examples and facts to inform readers. Let's take a look at the following "Cheetahs" passage to find examples of supporting details.

Cheetahs

Cheetahs are one of the fastest mammals on land, reaching up to 70 miles an hour over short distances. Even though cheetahs can run as fast as 70 miles an hour, they usually only have to run half that speed to catch up with their choice of prey. Cheetahs cannot maintain a fast pace over long periods of time because they will overheat their bodies. After a chase, cheetahs need to rest for approximately 30 minutes prior to eating or returning to any other activity.

In the example above, supporting details include:

- Cheetahs reach up to 70 miles per hour over short distances.
- They usually only have to run half that speed to catch up with their prey.
- Cheetahs will overheat their bodies if they exert a high speed over longer distances.
- Cheetahs need to rest for 30 minutes after a chase.

Look at the diagram below (applying the cheetah example) to help determine the hierarchy of topic, main idea, and supporting details.

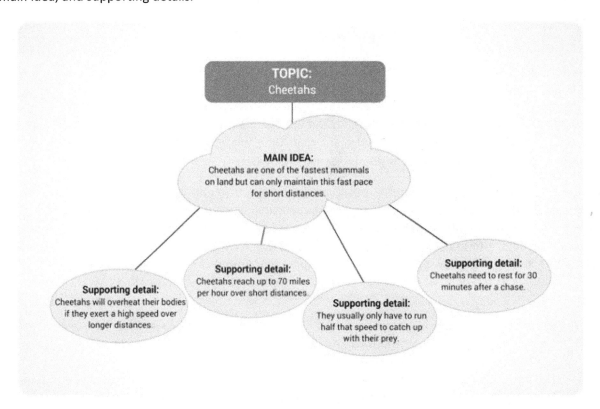

Inferences in a Text

Readers should be able to make *inferences*. Making an inference requires you to read between the lines and look for what is *implied* rather than what is directly stated. That is, using information that is known from the text, the reader is able to make a logical assumption about information that is *not* directly stated but is probably true. Read the following passage:

"Hey, do you wanna meet my new puppy?" Jonathan asked.

"Oh, I'm sorry but please don't—" Jacinta began to protest, but before she could finish, Jonathan had already opened the passenger side door of his car and a perfect white ball of fur came bouncing towards Jacinta.

"Isn't he the cutest?" beamed Jonathan.

"Yes—achoo!—he's pretty—aaaachooo!!—adora—aaa—aaaachoo!" Jacinta managed to say in between sneezes. "But if you don't mind, I—I—achoo!—need to go inside."

Which of the following can be inferred from Jacinta's reaction to the puppy?
    a. she hates animals
    b. she is allergic to dogs
    c. she prefers cats to dogs
    d. she is angry at Jonathan

An inference requires the reader to consider the information presented and then form their own idea about what is probably true. Based on the details in the passage, what is the best answer to the question? Important details to pay attention to include the tone of Jacinta's dialogue, which is overall polite and apologetic, as well as her reaction itself, which is a long string of sneezes. Answer choices (a) and (d) both express strong emotions ("hates" and "angry") that are not evident in Jacinta's speech or actions. Answer choice (c) mentions cats, but there is nothing in the passage to indicate Jacinta's feelings about cats. Answer choice (b), "she is allergic to dogs," is the most logical choice—based on the fact that she began sneezing as soon as a fluffy dog approached her, it makes sense to guess that Jacinta might be allergic to dogs. So even though Jacinta never directly states, "Sorry, I'm allergic to dogs!" using the clues in the passage, it is still reasonable to guess that this is true.

Making inferences is crucial for readers of literature, because literary texts often avoid presenting complete and direct information to readers about characters' thoughts or feelings, or they present this information in an unclear way, leaving it up to the reader to interpret clues given in the text. In order to make inferences while reading, readers should ask themselves:

- What details are being presented in the text?
- Is there any important information that seems to be missing?
- Based on the information that the author *does* include, what else is probably true?
- Is this inference reasonable based on what is already known?

Cite Specific Textual Evidence to Support Analysis of Primary and Secondary Sources

Primary and Secondary Sources

Primary sources are best defined as records or items that serve as evidence of periods of history. To be considered primary, the source documents or objects must have been created during the time period in which they reference. Examples include diaries, newspaper articles, speeches, government documents, photographs, and historical artifacts. In today's digital age, primary sources, which were once in print, often are embedded in secondary sources. Secondary sources, such as websites, history books, databases, or reviews, contain analysis or commentary on primary sources. Secondary sources borrow information from primary sources through the process of quoting, summarizing, or paraphrasing.

Analyzing and Citing a Text

When analyzing a text, readers first want to understand what type of text they are dealing with. For example, is the text a primary or secondary source? Is it meant to entertain or persuade? Who is the intended audience of the text? Once you answer these basic questions, it is critical to read the passage until the passage is completely understood. Once the passage is read, briefly summarize the text. Summarizing, or "translating," a text in one's own words, will help you to better comprehend the meaning and intent of the passage. Once you begin analyzing the passage in your own writing, the audience will expect you to draw upon the original passage for text evidence. This kind of evidence supports authorial credibility.

One way to draw from the passage read is to cite direct quotations from the text; this will show readers that you are intimate with the passage and that you are able to show them the original source from whence your information came. Including a paraphrase of the passage you draw from is also substantial, as long as you cite the author's last name with the page number. Depending on what style you are using (APA, MLA, or Chicago), references can be added at the end of your analysis to show exactly where the

source is from. References will tell you if the information is found in a magazine article, journal, or website, who the author is, and when the text was published.

Cite Specific Textual Evidence to Support Analysis of Science and Technical Texts

Backing up analysis as a science and technical text writer is a must if the end goal is to have a credible standing with an audience. Primary sources of science and technical writing draw upon observation and experiments to inform the reader; before science writing is published in a reputable journal, it must be peer reviewed. Therefore, a credible citation of a science or technical source in order to support your own analysis of the text is key in convincing your audience that the material you have is valid.

Keep in mind that the material you cite in science and technical writing must be precise. Precision writing in a science or technical field is imperative. Analysis of science writing does not need your opinion or a vague generalization on what the text says; it needs *precise* details from the text, detailing specific information regarding how the experiment or process works. Let's say you are writing in a science field and you want to explain to your audience how light travels. If you say "light travels fast," this is a vague generalization of complex material that should be treated thoroughly. The following is a more precise explanation of how light travels, and it draws upon a credible source from Stanford University:

> The first person to measure how light travels was astronomer Ole Roemer in 1675. He noticed that light rays from Jupiter were later than expected, and thus figured that light actually *travelled*, like we do, instead of being this immediate energy source. Today, the best understanding that we have of light is that it is "a disturbance in the electromagnetic fields of charged bodies" (Dr. Sten Odenwald, 1997). In a sense, light is energy that comes from a charge and projects outward into space, and it travels at 186,282 miles per second.

A few things to notice in the above explanation is a summary of information as well as a direct quotation. Writers should use direct quotations when something is said that they cannot say better themselves, or when there is no other way to say a truth or phrase. Notice how, after the direct quotation, there is the author's name followed by the year written in parenthesis. After you write your own analysis of a text in science or technical writing, it is important to ask the following questions: Have I explained the process in detail? Have I drawn upon the original source as an accurate reflection of the information? Always cite as many specifics from the original source as you can, while making sure to attribute the information to the proper source.

Determine a Theme or Central Idea of a Text

Main Idea or Primary Purpose

The main idea is what the writer wants to say about that topic. A writer may make the point that global warming is a growing problem that must be addressed in order to save the planet. Therefore, the topic is global warming, and the main idea is that it's *a serious problem needing to be addressed*. The topic can be expressed in a word or two, but the main idea should be a complete thought.

In order to illustrate the main idea, a writer will use supporting details—the details that provide evidence or examples to help make a point. Supporting details are typically found in nonfiction texts that seek to inform or persuade the reader.

For example, in the example of global warming, where the author's main idea is to show the seriousness of this growing problem and the need for change, the use of supporting details would be critical in effectively making that point. Supporting details used here might include statistics on an increase in global temperatures and studies showing the impact of global warming on the planet. The author could also include projections regarding future climate change in order to illustrate potential lasting effects of global warming.

Some questions may also ask you what the best title is for the passage. Going back to the *topic*, for these questions, it's important give a narrower answer that still encompasses the main idea of the passage. Asking for the appropriate title for passages is rare, but it's best to be prepared for anything.

Summarizing

At the end of a text or passage, it is important to summarize what the readers read. Summarizing is a strategy in which readers determine what is important throughout the text or passage, shorten those ideas, and rewrite or retell it in their own words. A summary should identify the main idea of the text or passage. Important details or supportive evidence should also be accurately reported in the summary. If writers provide irrelevant details in the summary, it may cloud the greater meaning of the summary in the text. When summarizing, writers should not include their opinions, quotes, or what they thought the author should have said. A clear summary provides clarity of the text or passage to the readers. Let's review the checklist of items writers should include in their summary.

Summary Checklist
- Title of the story
- Who is or are the main character(s)?
- What did the character(s) want?
- What was the problem?
- How did the character(s) solve the problem?
- How did the story end? What was the resolution?

Determine the central ideas or conclusions of a text

How Authors Develop Theme

Authors employ a variety of techniques to present a theme. They may compare or contrast characters, events, places, ideas, or historical or invented settings to speak thematically. They may use analogies, metaphors, similes, allusions, or other literary devices to convey the theme. An author's use of diction, syntax, and tone can also help convey the theme. Authors will often develop themes through the development of characters, use of the setting, repetition of ideas, use of symbols, and through contrasting value systems. Authors of both fiction and nonfiction genres will use a variety of these techniques to develop one or more themes.

Regardless of the literary genre, there are commonalities in how authors, playwrights, and poets develop themes or central ideas.

Authors often do research, the results of which contributes to theme. In prose fiction and drama, this research may include real historical information about the setting the author has chosen or include elements that make fictional characters, settings, and plots seem realistic to the reader. In nonfiction, research is critical since the information contained within this literature must be accurate and, moreover, accurately represented.

In fiction, authors present a narrative conflict that will contribute to the overall theme. In fiction, this conflict may involve the storyline itself and some trouble within characters that needs resolution. In nonfiction, this conflict may be an explanation or commentary on factual people and events.

Authors will sometimes use character motivation to convey theme, such as in the example from *Hamlet* regarding revenge. In fiction, the characters an author creates will think, speak, and act in ways that effectively convey the theme to readers. In nonfiction, the characters are factual, as in a biography, but authors pay particular attention to presenting those motivations to make them clear to readers.

Authors also use literary devices as a means of conveying theme. For example, the use of moon symbolism in Mary Shelley's *Frankenstein* is significant as its phases can be compared to the phases that the Creature undergoes as he struggles with his identity.

The selected point of view can also contribute to a work's theme. The use of first-person point of view in a fiction or non-fiction work engages the reader's response differently than third person point of view. The central idea or theme from a first-person narrative may differ from a third-person limited text.

In literary nonfiction, authors usually identify the purpose of their writing, which differs from fiction, where the general purpose is to entertain. The purpose of nonfiction is usually to inform, persuade, or entertain the audience. The stated purpose of a non-fiction text will drive how the central message or theme, if applicable, is presented.

Authors identify an audience for their writing, which is critical in shaping the theme of the work. For example, the audience for J.K. Rowling's *Harry Potter* series would be different than the audience for a biography of George Washington. The audience an author chooses to address is closely tied to the purpose of the work. The choice of an audience also drives the choice of language and level of diction an author uses. Ultimately, the intended audience determines the level to which that subject matter is presented and the complexity of the theme.

Paraphrasing

Another strategy readers can use to help them fully comprehend a text or passage is paraphrasing. Paraphrasing is when readers take the author's words and put them into their own words. When readers and writers paraphrase, they should avoid copying the text—that is plagiarism. It is also important to include as many details as possible when restating the facts. Not only will this help readers and writers recall information, but by putting the information into their own words, they demonstrate whether or not they fully comprehend the text or passage. Look at the example below showing an original text and how to paraphrase it.

Original Text: Fenway Park is home to the beloved Boston Red Sox. The stadium opened on April 20, 1912. The stadium currently seats over 37,000 fans, many of whom travel from all over the country to experience the iconic team and nostalgia of Fenway Park.

Paraphrased: On April 20, 1912, Fenway Park opened. Home to the Boston Red Sox, the stadium now seats over 37,000 fans. Many spectators travel to watch the Red Sox and experience the spirit of Fenway Park.

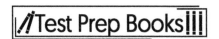

Paraphrasing, summarizing, and quoting can often cross paths with one another. Review the chart below showing the similarities and differences between the three strategies.

Paraphrasing	Summarizing	Quoting
Uses own words	Puts main ideas into own words	Uses words that are identical to text
References original source	References original source	Requires quotation marks
Uses own sentences	Shows important ideas of source	Uses author's own words and ideas

Analyze a Complex Set of Ideas or Sequence of Events

Logical Sequence

Even if the writer includes plenty of information to support their point, the writing is only coherent when the information is in a logical order. Logical sequencing is really just common sense, but it's an important writing technique. First, the writer should introduce the main idea, whether for a paragraph, a section, or the entire piece. Second, they should present evidence to support the main idea by using transitional language. This shows the reader how the information relates to the main idea and the sentences around it. The writer should then take time to interpret the information, making sure necessary connections are obvious to the reader. Finally, the writer can summarize the information in a closing section.

Though most writing follows the above pattern, it isn't a set rule. Sometimes writers change the order for effect. For example, the writer can begin with a surprising piece of supporting information to grab the reader's attention, and then transition to the main idea. Thus, if a passage doesn't follow the logical order, don't immediately assume it's wrong. However, most writing usually settles into a logical sequence after a nontraditional beginning.

As a reader, you are expected to pursue the author's logic regardless of the confusing transitions or complex content. For example, postmodern literary theory is all about deconstruction of texts. Deconstructing texts is picking apart a text in order to understand it better, or to understand it through an alternate lens or perspective. Some theorists who write about deconstruction, such as Jacques Derrida, do not follow traditional logic as a rule, because their intent is to interrupt traditional logic and to disturb the "supposed" order of processes and systems. This makes his writing extremely hard to understand. As readers, it is our job to go through the text, even a literary deconstructionist text, line by line, in order to understand it. Sometimes what helps with analyzing a complex text is to write your own observations when transitions occur from one subject to the next, and to mark them in the text. Or, after each paragraph or section, write a brief summary in your own words about what you just read. Drawing diagrams also helps, if the author is trying to explain a phenomenon or process that is not easily understood. Use all your faculties to follow the author and uncover the truth of what they are trying to relay to their audience.

Analyze in Detail a Series of Events Described In a Text

Causality

Causality is simply cause and effect; Event A causes Effect B to come into existence. Determining the strength of the relationship between the cause and effect will make evaluating the argument much easier. The more the events are directly related, the greater the causality.

Be wary of language that implies that there is direct causation when none actually exists. This trick usually involves *correlation*, which is a separate concept. *Correlating events* simply occur at the same time. One event does not cause or affect the other. Here's an example:

> Jeffrey Johnston is having the best year of his career. He hit twice as many home runs as last year and increased his batting average by one hundred points. He clearly must have wanted to earn more money in the offseason.

The author is unreasonably identifying an athlete's desire to enhance his value as the cause of his career year. It's illogical to say that some intangible increase in motivation is the reason why a player doubled his productivity. There could be any number of reasons why Jeffrey Johnston is playing much better. Maybe he was a young player who was still developing? Maybe a new coach fixed his batting swing? Maybe he is finally healthy for the first time in his career? There are a number of explanations as to why he improved. There is not enough evidence to suggest that desiring more money caused the spike in production. An addition of more information to the argument is:

> Jeffrey Johnston is having the best year of his career. He hit twice as many home runs as last year and increased his batting average by one hundred points. He clearly must have wanted to earn more money in the offseason. After all, he spent the whole offseason talking about how he increased his training regimen to put himself in a better position to earn more money as a free agent.

This argument adds more justification for attributing the desire for more money to the player's successful career year. This argument states that Jeffrey Johnston publicly discussed working harder to get a bigger contract. However, a strong causation is still not established. Some other causal agent could have been a much bigger reason for the bump in productivity. Another addition to the argument is:

> Jeffrey Johnston is having the best year of his career. He hit twice as many home runs as last year and increased his batting average by one hundred points. He clearly must have wanted to earn more money in the offseason. After all, he spent the whole offseason talking about how he increased his training regimen to put himself in a better position to earn more money as a free agent. Not to mention, he tattooed dollar signs on his forearms and painted his bat green.

The author has now included Jeffrey's tattoo and custom bat as evidence of his complete focus on money. This is the strongest causation since it has the most evidence directly related to the conclusion. According to the argument, Jeffrey knew he was a free agent before the season, he made public comments about working hard for better compensation, and he illustrated his mindset on his forearm and baseball bat. It seems likely that a desire for more money caused Jeffrey's career year.

Follow Precisely a Complex Multistep Procedure When Carrying Out Experiments

Interpreting Observed Data or Information

An important skill for scientific comprehension is the ability to interpret observed data and information. Scientific studies and research articles can be challenging to understand, but by familiarizing oneself with the format, language, and presentation of such information, fluency regarding the research process, results, and significance of such results can be improved. Most scientific research articles published in a journal follow the same basic format and contain the following sections:

- Abstract: A brief summary in paragraph form of each of the individual sections that are detailed below.

- Introduction or background: this section introduces the research question, states why the question is important, lays the groundwork for what is already known and what remains unknown, and states the hypothesis. Typically, this section includes a miniature literature review of previously published studies and details the results and or gaps in the research that these studies failed to achieve.

- Methods: this section details the experimental process that was conducted including information about how subjects were recruited, their demographic information, steps that were followed in the methodology, and what statistical analyses were performed on the data.

- Results: this section includes a narrative description of the results that were found as well as tables and graphs of the findings.

- Conclusion: this section interprets the results to put meaning to the numbers. Essentially, it answers the question, so what? It also discusses strengths, weaknesses, and shortcomings of the performed experiment as well as needed areas of future research.

Applying Scientific Principles

The scientific method provides the framework for studying and learning about the world in a scientific fashion. The scientific method has been around since at least the 17th century and is a codified way to answer natural science questions. Due to objectivity, the scientific method is impartial and its results are highly repeatable; these are its greatest advantages. There is no consensus as to the number of steps involved in executing the scientific method, but the following six steps are needed to fulfill the criteria for correct usage of the scientific method:

- Ask a question: Most scientific investigations begin with a question about a specific problem.

- Make observations: Observations will help pinpoint research objectives on the quest to answer the question.

- Create or propose a hypothesis: The hypothesis represents a possible solution to the problem. It is a simple statement predicting the outcome of an experiment testing the hypothesis.

- Formulate an experiment: The experiment tests the proposed hypothesis.

- Test the hypothesis: The outcome of the experiment to test the hypothesis is the most crucial step in the scientific method.

- Accept or reject the hypothesis: Using results from the experiment, a scientist can conclude to accept or reject the hypothesis.

Several key nuances of the scientific method include:

- The hypothesis must be verifiable and falsifiable. Falsifiable refers to the possibility of a negative solution to the hypothesis. The hypothesis should also have relevance, compatibility, testability, simplicity, and predictive power.

- Investigation must utilize both deductive and inductive reasoning. Deductive reasoning employs a logical process to arrive at a conclusion using premises considered true, while inductive reasoning employs an opposite approach. Inductive reasoning allows scientists to propose hypotheses in the scientific method, while deductive reasoning allows scientists to apply hypotheses to particular situations.

- An experiment should incorporate an independent, or changing, variable and a dependent, or non-changing, variable. It should also utilize both a control group and an experimental group. The experimental group will ultimately be compared against the control group.

A scientific explanation has three crucial components—a claim, evidence, and logical reasoning. A claim makes an assertion or conclusion focusing on the original question or problem. The evidence provides backing for the claim and is usually in the form of scientific data. The scientific data must be appropriate and sufficient. The scientific reasoning connects the claim and evidence and explains why the evidence supports the claim.

Scientific explanations must fit certain criteria and be supported by logic and evidence. The following represent scientific explanation criteria. The proposed explanation:

- Must be logically consistent
- Must abide by the rules of evidence
- Must report procedures and methods
- Must be open to questions and possible modification
- Must be based on historical and current scientific knowledge

The scientific method encourages the growth and communication of new information and procedures among scientists. Explanations of how the natural world shifts based on fiction, personal convictions, religious morals, mystical influences, superstitions, or authorities are not scientific and therefore irrelevant.

Scientific explanations have two fundamental characteristics. First, they should explain all scientific data and observations gleaned from experiments. Second, they should allow for predictions that can be verified with future experiments.

Craft and Structure

Determine the Meaning of Words and Phrases as They are Used in the Text

Meaning of Words in Context

There will be many occasions in one's reading career in which an unknown word or a word with multiple meanings will pop up. There are ways of determining what these words or phrases mean that do not require the use of the dictionary, which is especially helpful during a test where one may not be available. Even outside of the exam, knowing how to derive an understanding of a word via context clues will be a critical skill in the real world. The context is the circumstances in which a story or a passage is happening and can usually be found in the series of words directly before or directly after the word or phrase in question. The clues are the words that hint towards the meaning of the unknown word or phrase.

There may be questions that ask about the meaning of a particular word or phrase within a passage. There are a couple ways to approach these kinds of questions:

1. Define the word or phrase in a way that is easy to comprehend (using context clues).
2. Try out each answer choice in place of the word.

To demonstrate, here's an example from *Alice in Wonderland*:

> Alice was beginning to get very tired of sitting by her sister on the bank, and of having nothing to do: once or twice she <u>peeped</u> into the book her sister was reading, but it had no pictures or conversations in it, "and what is the use of a book," thought Alice, "without pictures or conversations?"

Q: As it is used in the selection, the word <u>peeped</u> means:

Using the first technique, before looking at the answers, define the word "peeped" using context clues and then find the matching answer. Then, analyze the entire passage in order to determine the meaning, not just the surrounding words.

To begin, imagine a blank where the word should be and put a synonym or definition there: "once or twice she _____ into the book her sister was reading." The context clue here is the book. It may be tempting to put "read" where the blank is, but notice the preposition word, "into." One does not read *into* a book, one simply reads a book, and since reading a book requires that it is seen with a pair of eyes, then "look" would make the most sense to put into the blank: "once or twice she <u>looked </u>into the book her sister was reading."

Once an easy-to-understand word or synonym has been supplanted, readers should check to make sure it makes sense with the rest of the passage. What happened after she looked into the book? She thought to herself how a book without pictures or conversations is useless. This situation in its entirety makes sense.

Now check the answer choices for a match:
    a. To make a high-pitched cry
    b. To smack
    c. To look curiously
    d. To pout

Since the word was already defined, Choice C is the best option.

Using the second technique, replace the figurative blank with each of the answer choices and determine which one is the most appropriate. Remember to look further into the passage to clarify that they work, because they could still make sense out of context.

a. Once or twice she <u>made a high pitched cry</u> into the book her sister was reading
b. Once or twice she <u>smacked</u> into the book her sister was reading
c. Once or twice she <u>looked curiously</u> into the book her sister was reading
d. Once or twice she <u>pouted</u> into the book her sister was reading

For Choice A, it does not make much sense in any context for a person to yell into a book, unless maybe something terrible has happened in the story. Given that afterward Alice thinks to herself how useless a book without pictures is, this option does not make sense within context.

For Choice B, smacking a book someone is reading may make sense if the rest of the passage indicates a reason for doing so. If Alice was angry or her sister had shoved it in her face, then maybe smacking the book would make sense within context. However, since whatever she does with the book causes her to think, "what is the use of a book without pictures or conversations?" then answer Choice B is not an appropriate answer. Answer Choice C fits well within context, given her subsequent thoughts on the matter. Answer Choice D does not make sense in context or grammatically, as people do not "pout into" things.

This is a simple example to illustrate the techniques outlined above. There may, however, be a question in which all of the definitions are correct and also make sense out of context, in which the appropriate context clues will really need to be honed in on in order to determine the correct answer. For example, here is another passage from *Alice in Wonderland*:

> . . . but when the Rabbit actually took a watch out of its waistcoat pocket, and looked at it, and then hurried on, Alice <u>started</u> to her feet, for it flashed across her mind that she had never before seen a rabbit with either a waistcoat-pocket or a watch to take out of it, and burning with curiosity, she ran across the field after it, and was just in time to see it pop down a large rabbit-hole under the hedge.

Q: As it is used in the passage, the word started means
a. To turn on
b. To begin
c. To move quickly
d. To be surprised

All of these words qualify as a definition of "start," but using context clues, the correct answer can be identified using one of the two techniques above. It's easy to see that one does not turn on, begin, or be surprised to one's feet. The selection also states that she "ran across the field after it," indicating that she was in a hurry. Therefore, to move quickly would make the most sense in this context.

The same strategies can be applied to vocabulary that may be completely unfamiliar. In this case, focus on the words before or after the unknown word in order to determine its definition. Take this sentence, for example:

> Sam was such a <u>miser</u> that he forced Andrew to pay him twelve cents for the candy, even though he had a large inheritance and he knew his friend was poor.

Unlike with assertion questions, for vocabulary questions, it may be necessary to apply some critical thinking skills that may not be explicitly stated within the passage. Think about the implications of the passage, or what the text is trying to say. With this example, it is important to realize that it is considered unusually stingy for a person to demand so little money from someone instead of just letting their friend have the candy, especially if this person is already wealthy. Hence, a <u>miser</u> is a greedy or stingy individual.

Questions about complex vocabulary may not be explicitly asked, but this is a useful skill to know. If there is an unfamiliar word while reading a passage and its definition goes unknown, it is possible to miss out on a critical message that could inhibit the ability to appropriately answer the questions. Practicing this technique in daily life will sharpen this ability to derive meanings from context clues with ease.

Determine the Meaning of Words and Phrases as They are Used in a Text

The previous section looks at meaning of word choice in literary text. In addition to literary text, you are also expected to know how to find out the meaning of word choice in informational text. Remember that in any kind of writing, it is important to look at the context surrounding the word in question. Informational texts include the following:

- Science articles
- History books
- Magazines
- Autobiographies
- Instruction Manuals
- Newspaper
- Court Opinion

Let's look at an example of how to look at the meaning of words in technical documents and how they affect tone. The following passage explains what lethal force is and can be found in a journal on crime reports:

> Lethal force, or deadly force, is defined as the physical means to cause death or serious harm to another individual. The law holds that lethal force is only accepted when you or another person are in immediate and unavoidable danger of death or severe bodily harm. For example, a person could be beating a weaker person in such a way that they are suffering severe enough trauma that could result in death or serious harm. This would be an instance where lethal force would be acceptable and possibly the only way to save that person from irrevocable damage.

Now let's look at this passage from a travel agency:

> Vacationers looking for a perfect experience should opt out of Disney parks and try a trip on Disney Cruise Lines. While a park offers rides, characters, and show experiences, it also includes long lines, often very hot weather, and enormous crowds. A Disney Cruise, on the other hand, is a relaxing, luxurious vacation that includes many of the same experiences as the parks, minus the crowds and lines. The cruise has top-notch food, maid service, water slides, multiple pools, Broadway-quality shows, and daily character experiences for kids. There are also many activities, such as bingo, trivia contests, and dance parties that can entertain guests of all ages. The cruise even stops at Disney's private island for a beach barbecue with characters,

waterslides, and water sports. Those looking for the Disney experience without the hassle should book a Disney cruise.

What is the tone of each passage and how does the author's word choice affect that tone? Let's look at the first passage and the word choice they use. They "define" lethal force in the first sentence and refer to people as "individuals." Right off we know that we are dealing with formal language, and this creates a serious tone. We also know that the first passage does not wish to undermine the seriousness of their subject by words like "unavoidable danger of death" and "severe bodily harm." In contrast, let's look at the passage from the travel agency. Some of the more memorable words they use are "relaxing," "luxurious," and "perfect experience"—these words choices are fun and fancy, which creates a lighthearted, desirable tone.

Both of these are technical texts which are meant to inform the reader. The difference is in their word choice affecting their tone. The authors here paid close attention to who their audience was. The audience of the first passage is possibly a police officer or a judge reading about the definition of lethal force for a case or a court trial. The audience for the second passage is probably a family looking to have fun and go on vacation. Both passages identified their audience and constructed the tone accordingly.

Determine the Meaning of Symbols, Key Terms, and Other Domain-Specific Words and Phrases

Determining unknown symbols or terms in a scientific or technical textbook is much easier today than one would expect. Textbooks almost always have a chart to depict what various symbols mean throughout their pages. Textbooks also have key terms bolded or italicized to help readers find the unknown word either at the bottom of a page in a footnote or in the back of a textbook inside a glossary. Electronic textbooks will even have links to click that will take you to a definition source or for extra resources to find if the textbook information does not suffice.

For example, let's say you are asked to do a research project on earthquakes. In order to do a comprehensive project explaining what earthquakes are and what they do, you would have to learn the symbols used to measure the earthquakes and the tools used to locate them. We would learn that earthquakes are measured by the symbol MS, meaning the standard surface-wave formula. The MS number goes from a 6.0 to an 8.9. The text depicting the term "MS" would probably be in parenthesis after the words or in a footnote to explain what it stands for and how it measures the earthquake. The term "standard surface-wave" might be in italics or bold to highlight the importance of the word.

Whatever source you are looking into for information, if it is credible, such as a textbook, scientific journal, or a book written by a meteorologist with a PhD, it should have the information you are looking for within the text, as well as additional resources to navigate to, such as a hyperlink or an appendix, in order to guide readers to the best information possible. Don't ignore the information in the margins or in the footnotes—this will be your best key to unlock the information you are struggling to comprehend. Also, don't just glance over words or symbols you don't know. Chances are, they will be listed somewhere on that page with more information. If all else fails and there is no additional resources or information available for a symbol or key word, you can always use a search engine, such as google, to look up a definition or key word. Once you find a credible source online to read, you will be able to find the information you are looking for.

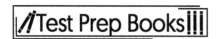

Analyze in Detail How an Author's Ideas or Claims are Developed and Refined

In non-fiction writing, authors employ argumentative techniques to present their opinion to readers in the most convincing way. First of all, persuasive writing usually includes at least one type of appeal: an appeal to logic (logos), emotion (pathos), or credibility and trustworthiness (ethos). When a writer appeals to logic, they are asking readers to agree with them based on research, evidence, and an established line of reasoning. An author's argument might also appeal to readers' emotions, perhaps by including personal stories and anecdotes (a short narrative of a specific event). A final type of appeal, appeal to authority, asks the reader to agree with the author's argument on the basis of their expertise or credentials. Consider three different approaches to arguing the same opinion:

Logic (Logos)
This is an example of an appeal to logic:

> Our school should abolish its current ban on cell phone use on campus. This rule was adopted last year as an attempt to reduce class disruptions and help students focus more on their lessons. However, since the rule was enacted, there has been no change in the number of disciplinary problems in class. Therefore, the rule is ineffective and should be done away with.

The author uses evidence to disprove the logic of the school's rule (the rule was supposed to reduce discipline problems; the number of problems has not been reduced; therefore, the rule is not working) and call for its repeal.

Emotion (Pathos)
An author's argument might also appeal to readers' emotions, perhaps by including personal stories and anecdotes. The next example presents an appeal to emotion. By sharing the personal anecdote of one student and speaking about emotional topics like family relationships, the author invokes the reader's empathy in asking them to reconsider the school rule.

> Our school should abolish its current ban on cell phone use on campus. If they aren't able to use their phones during the school day, many students feel isolated from their loved ones. For example, last semester, one student's grandmother had a heart attack in the morning. However, because he couldn't use his cell phone, the student didn't know about his grandmother's accident until the end of the day—when she had already passed away and it was too late to say goodbye. By preventing students from contacting their friends and family, our school is placing undue stress and anxiety on students.

Credibility (Ethos)
Finally, an appeal to authority includes a statement from a relevant expert. In this case, the author uses a doctor in the field of education to support the argument. All three examples begin from the same opinion—the school's phone ban needs to change—but rely on different argumentative styles to persuade the reader.

> Our school should abolish its current ban on cell phone use on campus. According to Dr. Bartholomew Everett, a leading educational expert, "Research studies show that cell phone usage has no real impact on student attentiveness. Rather, phones provide a valuable technological resource for learning. Schools need to learn how to integrate this new technology into their curriculum." Rather than banning phones altogether, our school should follow the advice of experts and allow students to use phones as part of their learning.

Analyze and Evaluate the Effectiveness of the Structure an Author Uses In His or Her Exposition or Argument

Evaluating the Effectiveness of a Piece of Writing

An effective and engaging piece of writing will cause the reader to forget about the author entirely. Readers will become so engrossed in the subject, argument, or story at hand that they will almost identify with it, readily adopting beliefs proposed by the author or accepting all elements of the story as believable. On the contrary, poorly written works will cause the reader to be hyperaware of the author, doubting the writer's knowledge of a subject or questioning the validity of a narrative. Persuasive or expository works that are poorly researched will have this effect, as well as poorly planned stories with significant plot holes. An author must consider the task, purpose, and audience to sculpt a piece of writing effectively.

When evaluating the effectiveness of a piece, the most important thing to consider is how well the purpose is conveyed to the audience through the mode, use of rhetoric, and writing style.

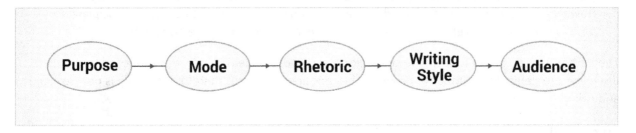

The purpose must pass through these three aspects for effective delivery to the audience. If any elements are not properly considered, the reader will be overly aware of the author, and the message will be lost. The following is a checklist for evaluating the effectiveness of a piece:

- Does the writer choose the appropriate writing mode—argumentative, narrative, descriptive, informative—for his or her purpose?
- Does the writing mode employed contain characteristics inherent to that mode?
- Does the writer consider the personalities/interests/demographics of the intended audience when choosing rhetorical appeals?
- Does the writer use appropriate vocabulary, sentence structure, voice, and tone for the audience demographic?
- Does the author properly establish himself/herself as having authority on the subject, if applicable?
- Does the piece make sense?

Another thing to consider is the medium in which the piece was written. If the medium is a blog, diary, or personal letter, the author may adopt a more casual stance towards the audience. If the piece of writing is a story in a book, a business letter or report, or a published article in a journal or if the task is to gain money or support or to get published, the author may adopt a more formal stance. Ultimately, the writer will want to be very careful in how he or she addresses the reader.

Finally, the effectiveness of a piece can be evaluated by asking how well the purpose was achieved. For example, if students are assigned to read a persuasive essay, instructors can ask whether the author influences students' opinions. Students may be assigned two differing persuasive texts with opposing perspectives and be asked which writer was more convincing. Students can then evaluate what factors

contributed to this—for example, whether one author uses more credible supporting facts, appeals more effectively to readers' emotions, presents more believable personal anecdotes, or offers stronger counterargument refutation. Students can then use these evaluations to strengthen their own writing skills.

Analyze a Particular Point of View or Cultural Experience Reflected in a Work of Literature from Outside the United States

Developing a knowledge of diverse texts other than those in western literature is an important part of learning about other cultures and their point of view. Since literature reflects the time period, consciousness, and perspectives of a culture, reading diverse texts in alternate time periods is a fundamental way to experience cultures other than our own. We might not only learn compassion but a well-rounded view of life and experiences around the world. The following list depicts world literature classics from other cultures, although the list is not comprehensive:

- *Don Quixote* by Miguel de Cervantes
- *Things Fall Apart* by Chinua Achebe
- *Pride and Prejudice* by Jane Austen
- *Wuthering Heights* by Emily Bronte
- *Heart of Darkness* by Joseph Conrad
- *The Stranger* by Albert Camus
- *The Divine Comedy* by Dante Alighieri
- *Canterbury Tales* by Geoffrey Chaucer
- *Great Expectations* by Charles Dickens
- *Madame Bovary* by Gustave Flaubert
- *One Hundred Years of Solitude* by Gabriel Garcia Marquez
- *The Iliad* and *The Odyssey* by Homer
- *The Trial* by Franz Kafka
- *The Sound of the Mountain* by Yasunari Kawabata
- *Diary of a Madman and Other Stories* by Lun Xun
- *The Tale of Genji* by Murasaki Shikibu
- *Lolita* by Vladimir Nabokov
- *The Book of Disquiet* by Fernando Pessoa
- *War and Peace* by Leo Tolstoy
- *Mrs. Dalloway* by Virginia Woolf
- *Nectar in a Sieve* by Kamala Markandaya
- *Obasan* by Joy Kogawa
- *The House of the Spirits* by Isabel Allende

Let's look at an example of a world literature classic and its surrounding context. The following is a passage from the narrative poem, The *Divine Comedy*, written by Dante Alighieri, an Italian poet in the late Middle ages.

> Benign Apollo! this last labour aid,
> And make me such a vessel of thy worth,
> As thy own laurel claims of me belov'd.
> Thus far hath one of steep Parnassus' brows
> Suffic'd me; henceforth there is need of both

For my remaining enterprise Do thou
Enter into my bosom, and there breathe
So, as when Marsyas by thy hand was dragg'd
Forth from his limbs unsheath'd. O power divine!
If thou to me of shine impart so much,
That of that happy realm the shadow'd form
Trac'd in my thoughts I may set forth to view,
Thou shalt behold me of thy favour'd tree
Come to the foot, and crown myself with leaves;
For to that honour thou, and my high theme
Will fit me. If but seldom, might Sire!
To grace his triumph gathers thence a wreath
Caesar or bard (more shame for human wills
Deprav'd!) joy to the Delphic god must spring
From the Pierian foliage, when one breast
Is with such thirst inspir'd. From a small spark
Great flame hath risen; after me perchance
Others with better voice may pray, and gain
From the Cirrhaean city answer kind.

This text was written in the Middle Ages, as you can probably tell from the language using "thy" and "thou." Remember that most texts in world literature from other countries that you read in English are probably translated into English. This poem was originally written in Italian. The original poem in Italian has a different rhyme scheme and metrical line than the one above. The above is iambic pentameter, which is a popular line for a poet to write in in English because it fits nicely with the way we stress our words. In the original Italian, the poem is written in terza rima, which is a rhyme scheme that fits well with the Italian language because many of the word-endings in Italian are easier to rhyme than in English. In the poem above, the poet is calling on Apollo for inspiration, who is the god of poetry. This rhetorical device of calling out to an abstract thing is called an apostrophe and is used by many poets throughout the centuries, though this device is less common now. In epic poems or medieval narrative poems, it was very popular for the author to dedicate a passage to the gods before embarking on the journey, or narrative. Knowing characteristics and context of world literature can help us to understand cultures better in the past and why they did certain things.

Analyze a Case in Which Grasping Point Of View Requires Distinguishing What Is Directly Stated in a Text from What Is Really Meant

Satire

Satire is a genre of literature that is humorous on the surface but deals with social commentary on a deeper level. Satire uses irony and sarcasm to deal with its subject and usually seeks to shame or shed light on a society that is corrupt or acting on irrational principles. The most famous example of satire is Jonathan Swift's "A Modest Proposal," an essay that sheds light on England's treatment of the Irish at the time. The essay proposes that poor Irish parents sell their children as food to the wealthy to help with economic circumstances, thus addressing harsh attitudes toward the poor. Satire is a genre that shows widely accepted behavior in a new perspective and works at changing social consciousness.

Some modern examples of satire are late night TV shows, such as *The Late Show* with John Colbert or *Saturday Night Live*, horror movies, or political cartoons. Many modern satire mediums deal with

political figures and discourse and the policies that affect the public. Horror movies often deal with self-parody, such as *Scream*, which is a horror movie that makes fun of other horror movies in American culture. Although it is sometimes harsh, satire can be an intelligent way for a society to laugh at itself even under grim circumstances.

Sarcasm

Depending on the tone of voice or the words used, sarcasm can be expressed in many different ways. Sarcasm is defined as a bitter or ambiguous declaration that intends to cut or taunt. Most of the ways we use sarcasm is saying something and not really meaning it. In a way, sarcasm is a contradiction that is understood by both the speaker and the listener to convey the opposite meaning. For example, let's say Bobby is struggling to learn how to play the trumpet. His sister, Gloria, walks in and tells him: "What a great trumpet player you've become!" This is a sort of verbal irony known as sarcasm. Gloria is speaking a contradiction, but Bobby and Gloria both know the truth behind what she's saying: that Bobby is not a good trumpet player. Sarcasm can also be accompanied by nonverbal language, such as a smirk or a head tilt. Remember that sarcasm is not always clear to the listener; sometimes sarcasm can be expressed by the speaker but lost on the listener.

Irony

Irony is a device that authors use when pitting two contrasting items or ideas against each other in order to create an effect. It's frequently used when an author wants to employ humor or convey a sarcastic tone. Additionally, it's often used in fictional works to build tension between characters, or between a particular character and the reader. An author may use *verbal irony* (sarcasm), *situational irony* (where actions or events have the opposite effect than what's expected), and *dramatic irony* (where the reader knows something a character does not). Examples of irony include:

- Dramatic Irony: An author describing the presence of a hidden killer in a murder mystery, unbeknownst to the characters but known to the reader.

- Situational Irony: An author relating the tale of a fire captain who loses her home in a five-alarm conflagration.

- Verbal Irony: This is where an author or character says one thing but means another. For example, telling a police officer "Thanks a lot" after receiving a ticket.

Understatement

Making an understatement means making a statement that gives the illusion of something being smaller than it actually is. Understatement is used, in some instances, as a humorous rhetorical device. Let's say that there are two friends. One of the friends, Kate, meets the other friend's, Jasmine's, boyfriend. Jasmine's boyfriend, in Kate's opinion, is attractive, funny, and intelligent. After Kate meets her friend's boyfriend, Kate says to Jasmine, "You could do worse." Kate and Jasmine both know from Kate's tone that this means Kate is being ironic—Jasmine could do much, much worse, because her boyfriend is considered a "good catch." The understatement was a rhetorical device used by Kate to let Jasmine know she approves.

Determine an Author's Point of View or Purpose in a Text

Identifying the Author's Purpose

Every story has a *narrator*, or someone who tells the story. This is sometimes also referred to as the story's *point of view*. Don't make the mistake of assuming that the narrator and the author are the same person—on the contrary, they may have completely different opinions and personalities. There are also several types of narrators commonly found in fiction, and each type serves a different purpose.

The first type of narrator is a *first-person narrator* who tells the story from an "I" perspective. A first-person narrator is almost always a character in the story. They might be a main protagonist (like Jane Eyre in the novel of the same name) or a side character (such as Nelly in *Wuthering Heights*). Also, first-person narrator has an immediate response to the events of the story. However, keep in mind that they are reporting the story through their own perspective, so this narrator may not be totally objective or reliable. Pay attention to language that shows the narrator's tone and consider how the events of the story might be interpreted differently by another character.

Another type of narrator is a *second-person narrator* who tells the story from a "you" perspective. Readers probably won't encounter this type of narrator often, although it's becoming slightly more common in modern literature. A second-person narrator forces the reader to insert themselves into the story and consider how they would respond to the action themselves.

Much more common is a *third-person narrator*. This type of narrator is further divided into two subtypes: *third-person limited* and *third-person omniscient*. In both cases, the story is told from an outside perspective ("he," "she," and "they," rather than "I," "me," or "you"). In this way, a third-person narrator is more distanced and objective than a first-person narrator. A third-person limited narrator tends to follow one main character in the story, reporting only that character's thoughts and only events that directly involve that character (think of the Harry Potter books, which are told by a third-person narrator but primarily limited to Harry's thoughts and experiences). A third-person omniscient narrator isn't limited to one character, but instead reports freely on the thoughts, actions, and events of all characters and situations in the story (omniscient means "all knowing"). Third-person omniscient is generally the most flexible and objective type of narrator. Readers can find this narrator in the novels of writers like Charles Dickens and Alexandre Dumas, whose stories are spread across dozens of different characters and locations.

By identifying the type of narrator, readers can assess a narrator's objectivity as well as how the narrator's perspective is limited to a single character's experience. When a narrator is unbiased, readers can take their account of events at face value; when a narrator is less objective, readers have to consider how the narrator's personal perspective influences the reader's understanding.

In addition to the point of view, another important factor to consider is the writer's *purpose*. This answers the question, "Why did the author write this?" Three of the most common purposes are to persuade, to inform, or to entertain. Of course, it's possible for these purposes to overlap.

Generally, literary texts are written to *entertain*. That is, the reader is simply supposed to enjoy the story! The romantic, dramatic, and adventurous elements are all designed to get readers involved with the fates of the characters. When the primary purpose of a text is to entertain, the focus is on the plot elements, the characters' thoughts and emotions, and descriptions of places, events, and characters. Examples of texts that entertain include novels, plays, poetry, and memoirs. Within the more general

purpose of entertaining, literary texts might include passages to convey emotion or to describe a character's thoughts or feelings.

Another common writing purpose is to *inform*. Informative writing intends to teach the reader about a particular topic. Generally, this type of writing is objective and unbiased. Informative writing may also be descriptive. However, instead of describing thoughts and emotions, informative writing tends to describe concrete facts. After reading an informative text, the reader should be more knowledgeable about the text's subject. Examples of informative writing include textbooks, research articles, and texts about academic topics like science and history. Sometimes, informative writing is also described as writing that intends to teach or give information about a subject.

The third common writing purpose is to *persuade*. Unlike the objective writing found in informative texts, persuasive texts are more subjective. Rather than being neutral, the author expresses an opinion about the subject and tries to convince readers to agree. As with informative texts, persuasive texts often include facts and statistics—however, these are used to support the writer's perspective. Examples of persuasive writing include newspaper editorials and texts about controversial topics like politics or social issues. Sometimes, persuasive writing is also described as writing that intends to convince, argue, or express an opinion about a subject.

In order to determine the purpose of a text, readers can keep the following questions in mind:

- What type of details are included (facts, emotions, imagery, etc.)?

- Facts and statistics tend to indicate informative or persuasive writing, while emotions or imagery point to writing that entertains.

- Is the author neutral or opinionated?

- An opinionated or biased author is probably writing to persuade, while a neutral author writes to inform.

- Is this passage teaching information?

- If readers gained a lot of new knowledge about a subject after reading a text, the primary purpose was probably to inform.

Compare the Point of View of Two or More Authors for How They Treat the Same or Similar Topics

At some point in your life you will be asked to compare two different things. Comparing two different texts can be difficult because it is tedious—you are being asked to read each text, formulate an analysis of each, and then compare the differences between the two texts. There are many instances where you might run into this in everyday life. Let's say you are being asked to compare two separate proposals for your workplace and you have to give a report on the characteristics of each. Some details you will look at are rhetorical devices used, effectiveness of the language, organization, style, and content, among others. Although the subject of the text is important, you will inevitably look at how each author connects to their audience and how *well* they do so.

The following two passages are written about seat belts. For the sake of discussion afterwards, let's pretend that they are written for an oral presentation for a college class. After you read each of them carefully, we will analyze the impact of each:

Passage A

Seat belts save more lives than any other automobile safety feature. Many studies show that airbags save lives as well; however, not all cars have airbags. For instance, some older cars don't. Furthermore, air bags aren't entirely reliable. For example, studies show that in 15% of accidents, airbags don't deploy as designed; but, on the other hand, seat belt malfunctions are extremely rare. The number of highway fatalities has plummeted since laws requiring seat belt usage were enacted.

Passage B

You are probably wondering about seat belts. A lot of studies show that seat belts help with car accidents. My friend got in a car accident one time and the seat belt saved him, although it left bruises on his chest. It took weeks for the bruises to go away. His name is Chris. But yes it did save his life. You should wear a seat belt every single time you get in a car because some cars do not have airbags or sometimes the airbags do not work and you never know when that will happen.

Let's talk about the overall difference between them. Which one is more convincing? Which one is better suited to fit its audience? After reading each, Passage A fits a college audience better than Passage B. It "flows" better and doesn't feel so awkward. But why?

First, let's look at point of view. Passage A sticks with a consistent perspective of a neutral third-person account. There are no "you's" or I's." Passage B, on the other hand, inserts second-person "you" in its passage and intermixes it with a neutral third person. When choosing a point of view, it's important to be consistent with one or the other. To personalize a text, sometimes "you" or "I" is used to create a familiarity with the readers. However, in a passage as short as this that's trying to convince its audience of something, third person is the best option.

Next, let's look at the evidence used in each of these passages. Passage A uses concrete evidence with data-driven statistics. We see that in "15% of accidents, airbags don't deploy as designed" in Passage A, but in Passage B, we receive information like this: "A lot of studies show that seat belts help with car accidents." The phrase "a lot" is vague, but more importantly, what do they mean by "seat belts help with car accidents"? *How* do they help with car accidents? The author of Passage B, maybe unintentionally, withholds important information from its audience. Clear evidence is very important to the credibility of the author. While Passage A has straightforward claims ("Seat belts save more lives than any other automobile safety feature"), Passage B uses vague language to hedge around the information they are providing.

Finally, the difference in organization between the two passages is important. Passage A begins with a straightforward claim, offers a counterclaim, and then proceeds to provide evidence against the counterclaim, thus proving the claim by the end of the passage. The organization of Passage B is a little more confusing. The author does not begin with a claim, but begins with an assumption that the audience "is wondering about seatbelts." Then the author provides a personal testimony, which is fine, but proceeds to get off track with the name of his friend and how long it took for his bruises to go away,

which is not part of the author's claim that "seat belts help with car accidents." The author concludes the claim but gets off track in the body of the passage. Thus, Passage A is better organized than Passage B.

Integration of Knowledge and Ideas

Delineate and Evaluate the Argument and Specific Claims in a Text

In order to evaluate the claims made in a text, it's important to look at the evidence offered to support those claims. Knowing whether an argument is faulty or valid will put you in a better starting point. Claims that are supported will usually have concrete evidence to back it up, like our seatbelt example in the previous section. Claims that are not supported will have generalizations, like "everyone who wears a seatbelt is safe." The word "everyone" is an absolute, and avoiding absolutes like "always" or "all" or "everyone" is important for the validity of a claim. Let's look at the flaws authors can make while presenting arguments:

Red Herring

A *red herring* is a logical fallacy in which irrelevant information is introduced to alter the argument's trajectory. Red herrings are the irrelevant information used to fallaciously and slyly divert the argument into an unrelated topic. This fallacy is common in thriller movies or television shows in which the audience is led to believe that a character is the villain or mastermind, while the true villain remains a secret. In terms of the Logical Reasoning section, red herrings will attempt to distract the reader with irrelevant information in either the question or answer choice. A red herring is sometimes referred to as a *straw man*, since this fallacy attacks a different argument than the one presented. Consider the following example of a red herring:

> The government must immediately issue tax cuts to strengthen the economy. A strong middle class is the backbone of a fully functional economy, and tax cuts will increase the discretionary spending necessary to support the middle class. After all, it's extremely important for our society to be open-minded and to limit racial discord.

On its face, the argument looks fine. The conclusion is obvious—the government needs to pass tax cuts to strengthen the economy. This is based on the premise that the cuts will increase discretionary spending, which will strengthen the middle class, and the economy will be stronger. However, the last sentence is a red herring. The argument does not address how a society should be open-minded and avoid racism; it holds no connection with the main thrust of the argument. In this scenario, look out for answer choices that address the red herring rather than the essential argument.

Extreme Language

The test makers commonly write appealing answer choices, but take the language to such an unjustified extreme that it is rendered incorrect. The extreme language usually will take the argument too far.

An example of an argument appealing to the extreme is:

> Weight lifting breaks down muscles and rebuilds them. If one just kept exercising and never stopped, his or her body would deteriorate and eventually fall apart.

This argument is clearly illogical. The author correctly describes what weight lifting does to the body, but then takes the argument to an unjustified extreme. If someone continually lifted weights, his or her

31

body would not deteriorate and fall apart due to the weights breaking down muscle. The weightlifter may eventually die from a heart attack or dehydration, but it would not be because of how weight lifting rebuilds muscle.

Appeals to Authority Fallacy

Arguing from authority occurs when an author uses an expert to justify their argument. Whether the appeal is fallacious depends on the status, authority, or expertise of the cited authority. If the authority cannot reasonably be relied upon, then the author is committing a logical fallacy.

Always keep in mind that *appealing to authority* can be valid or invalid argumentation. It all depends on the authority's credibility. How is the credibility of a cited authority evaluated? Indicators are listed below:

- Whatever expertise is stated in the field. The required qualifications will depend on the field, such as a PhD in chemistry or two decades experience as the top bricklayer.

- Whether the claim being asserted by the qualified authority is actually within his or her field of expertise.

- Whether the majority of other similarly qualified experts agree, or if there is open disagreement on the subject.

- Whether the author is biased in some manner.

- Whether the claimed field of expertise is legitimate. For example, be cautious of claimed expertise in psychic abilities or horoscope readings.

- Whether the authority is identified. The author will sometimes preface the argument with phrases like, *experts say*, *a book stated*, *a documentary reported*, and *they say*. The unidentified authority is almost always a sign of fallacious argumentation.

Always ask yourself these questions when an argument cites an authority. They will help you determine if the authority is reliable and relevant to the argument at hand. If the authority can be trusted, this may lend weight to one of the answer choices. If, however, the authority is in question, this can divert you to a better option. Here's an example of a flawed argument from authority:

> Bill Gates is the preeminent mathematical and computer genius of his era. Bill Gates believes in our government making huge economic investments in the research of renewable energy. The government should obviously follow this advice.

The argument relates Bill Gates' genius in math and computers. The author is hoping the inference is that math and computer genius translates to expertise in government fiscal and energy policy. The argument offers no support as to why the expertise would apply to this seemingly unrelated field. This is flawed reasoning. Although Bill Gates is an authority on computers, it does not mean that the government should follow his thoughts in an unrelated field.

Hasty Generalizations

A hasty generalization involves an argument relying on insufficient statistical data or inaccurately generalizing. One common generalization occurs when a group of individuals under observation have

some quality or attribute that is asserted to be universal or true for a much larger number of people than actually documented. Here's an example of a hasty generalization:

A man smokes a lot of cigarettes, but so did his grandfather. The grandfather smoked nearly two packs per day since his World War II service until he died at ninety years of age. Continuing to smoke cigarettes will clearly not impact the grandson's long-term health.

This argument is a hasty generalization because it assumes that one person's addiction and lack of consequences will naturally be reflected in a different individual. There is no reasonable justification for such extrapolation. It is common knowledge that any smoking is detrimental to one's health. The fact that the man's grandfather smoked two packs per day and lived a long life has no logical connection with the grandson engaging in similar behavior. The hasty generalization doesn't take into account other reasons behind the grandfather's longevity. Nor does the author offer evidence that might support the idea that the man would share a similar lifetime if he smokes. It might be different if the author stated that the man's family shares some genetic trait rendering them immune to the effects of tar and chemicals on the lungs. If this were in the argument, we would assume it as truth and find the generalization to be valid rather than hasty. Of course, this is not the case in our example.

COMMON LOGICAL FALLACIES		
Fallacy	**Summary**	**Example**
Red Herring	A red herring is a logical fallacy in which irrelevant information is introduced to alter the argument's trajectory and divert the argument into an unrelated topic.	The government must immediately issue tax cuts to strengthen the economy. A strong middle class is the backbone of a fully functional economy, and tax cuts will increase the discretionary spending necessary to support the middle class. After all, it's extremely important for our society to be open-minded and limit racial discord.
Extreme Language	Extreme arguments will take the language to such an unjustified extreme, or push it so far, that the extremeness will render the argument incorrect.	Weight lifting breaks down muscles and rebuilds them. If one just kept exercising and never stopped, their body would deteriorate and eventually fall apart.
Irrelevancy	Although usually persuasive, irrelevant information is unconnected to the argument's logic or context.	Argument only discusses the balance of speed and accuracy and does not mention history. Answer Choice: Historically, news organizations always waited to verify information and sources, and the change is due to the advent of the Internet.
Similar Language	Answer choices deceptively bear similar language, like the exact same words or phrases, as the argument.	Argument: This leads to some analysts jumping the gun and predicting outcomes inaccurately. Answer Choice: Jumping the gun is the worst mistake any analyst could make.
Parallel Reasoning	Parallel reasoning will make otherwise illogical answer choices seem more credible by mimicking the argument's structure.	Argument only involves news analysts prioritizing speed and accuracy. Answer Choice: The worst sports analysts prioritize speed over accuracy when announcing trades and other roster moves.
Appeal to Authority	Arguing from authority occurs when an author uses an expert to justify his argument, which is illogical if the supposed authority lacks the required status or expertise.	Bill Gates is the pre-imminent mathematical and computer genius of his era. Bill Gates believes in our government making huge economic investments in the research of renewable energy. The government should obviously follow this advice.
Hasty Generalizations	A hasty generalization involves an argument relying on insufficient statistical data or inaccurately gen-eralizing.	I smoke a lot of cigarettes but so did my grandpa. He smoked nearly two packs per day from his World War II service until he died at ninety years of age. Continuing to smoke cigarettes will clearly not impact my long-term health.
Confusing Correlation with Causation	Arguments confuse correlation with causation by saying that one event caused another just because they occurred at the same time.	Jacob adopted a puppy last week. I saw him on Monday and he's lost fifty pounds since last summer. He looks really good. Adopting a puppy must be the secret to weight loss.

Confusing Correlation with Causation

Correlation means that two events occur at the same time; however, *causation* indicates that one event caused a separate event. It is a common logical fallacy for correlation to be confused with causation. Arguments confuse correlation with causation by stating that one event caused another just because they occurred at the same time.

When evaluating whether an argument's reasoning is confusing the terms causation and correlation, always identify the actual cause of the event. Is there another more likely or reasonable event that could have been the true cause? If there is no causation, then there is a logical error. Also, be mindful of the events' timing and relationship to each other. If the author is drawing undue attention to the fact that the events occurred close in time, then check for this fallacy. However, remain aware that it is always possible for events occurring at the same time to be the cause.

Below is an example of an argument confusing correlation with causation:

> Jacob adopted a puppy two months ago. I saw him on Monday, and he's lost fifty pounds since summer. He looks really good. Adopting a puppy must be the secret to his weight loss.

In this argument, the author is saying that he saw Jacob with a new puppy, and Jacob looked good; therefore, adopting the puppy caused Jacob to look better. This is clearly illogical. Jacob could have been on a serious juice cleanse. He could have started running or lifting weights. Although the events (adopting the puppy and losing weight) occurred at the same time, it does not necessarily mean that one caused the other.

This is often a judgment decision. The argument must justify the assumption of causation in some meaningful way. For example, in the sample argument it would not be fallacious if the author stated that Jacob adopted the puppy a while ago, and they often run together. This would not be a logical flaw since starting to exercise with a new dog could reasonably cause the weight loss and improved looks.

Practice Questions

Questions 1-6 are based upon the following passage:

This excerpt is an adaptation of Jonathan Swift's *Gulliver's Travels into Several Remote Nations of the World.*

My gentleness and good behaviour had gained so far on the emperor and his court, and indeed upon the army and people in general, that I began to conceive hopes of getting my liberty in a short time. I took all possible methods to cultivate this favourable disposition. The natives came, by degrees, to be less apprehensive of any danger from me. I would sometimes lie down, and let five or six of them dance on my hand; and at last the boys and girls would venture to come and play at hide-and-seek in my hair. I had now made a good progress in understanding and speaking the language. The emperor had a mind one day to entertain me with several of the country shows, wherein they exceed all nations I have known, both for dexterity and magnificence. I was diverted with none so much as that of the rope-dancers, performed upon a slender white thread, extended about two feet, and twelve inches from the ground. Upon which I shall desire liberty, with the reader's patience, to enlarge a little.

This diversion is only practised by those persons who are candidates for great employments, and high favour at court. They are trained in this art from their youth, and are not always of noble birth, or liberal education. When a great office is vacant, either by death or disgrace (which often happens,) five or six of those candidates petition the emperor to entertain his majesty and the court with a dance on the rope; and whoever jumps the highest, without falling, succeeds in the office. Very often the chief ministers themselves are commanded to show their skill, and to convince the emperor that they have not lost their faculty. Flimnap, the treasurer, is allowed to cut a caper on the straight rope, at least an inch higher than any other lord in the whole empire. I have seen him do the summerset several times together, upon a trencher fixed on a rope which is no thicker than a common packthread in England. My friend Reldresal, principal secretary for private affairs, is, in my opinion, if I am not partial, the second after the treasurer; the rest of the great officers are much upon a par.

1.
Part A
Which of the following statements best summarizes the central purpose of this text?
   a. Gulliver details his fondness for the archaic yet interesting practices of his captors.
   b. Gulliver conjectures about the intentions of the aristocratic sector of society.
   c. Gulliver becomes acquainted with the people and practices of his new surroundings.
   d. Gulliver's differences cause him to become penitent around new acquaintances.

Part B

Which detail from the article best supports the answer to Part A?

a. "I have seen him do the summerset several times together, upon a trencher fixed on a rope which is no thicker than a common packthread in England."

b. "Flimnap, the treasurer, is allowed to cut a caper on the straight rope, at least an inch higher than any other lord in the whole empire."

c. "I was diverted with none so much as that of the rope-dancers, performed upon a slender white thread, extended about two feet, and twelve inches from the ground."

d. "My gentleness and good behaviour had gained so far on the emperor and his court, and indeed upon the army and people in general, that I began to conceive hopes of getting my liberty in a short time."

2. What is the word *principal* referring to in the following text?

My friend Reldresal, principal secretary for private affairs, is, in my opinion, if I am not partial, the second after the treasurer; the rest of the great officers are much upon a par.

a. Primary or chief
b. An acolyte
c. An individual who provides nurturing
d. One in a subordinate position

3.

Part A

What can the reader infer from this passage?

I would sometimes lie down, and let five or six of them dance on my hand; and at last the boys and girls would venture to come and play at hide-and-seek in my hair.

a. The children tortured Gulliver.
b. Gulliver traveled because he wanted to meet new people.
c. Gulliver is considerably larger than the children who are playing around him.
d. Gulliver has a genuine love and enthusiasm for people of all sizes.

Part B

Which detail from the article best supports the answer to Part A?

a. "I took all possible methods to cultivate this favourable disposition."

b. "Upon which I shall desire liberty, with the reader's patience, to enlarge a little."

c. "This diversion is only practised by those persons who are candidates for great employments, and high favour at court."

d. "They are trained in this art from their youth, and are not always of noble birth, or liberal education."

4. What is the significance of the word *mind* in the following passage?

The emperor had a mind one day to entertain me with several of the country shows, wherein they exceed all nations I have known, both for dexterity and magnificence.

a. The ability to think
b. A collective vote
c. A definitive decision
d. A mythological question

5. Which of the following assertions does not support the fact that games are a commonplace event in this culture?

a. My gentlest and good behavior . . . short time.

b. They are trained in this art from their youth . . . liberal education.

c. Very often the chief ministers themselves are commanded to show their skill . . . not lost their faculty.

d. Flimnap, the treasurer, is allowed to cut a caper on the straight rope . . . higher than any other lord in the whole empire.

6.

Part A

How do the roles of Flimnap and Reldresal serve as evidence of the community's emphasis in regards to the correlation between physical strength and leadership abilities?

a. Only children used Gulliver's hands as a playground.

b. The two men who exhibited superior abilities held prominent positions in the community.

c. Only common townspeople, not leaders, walk the straight rope.

d. No one could jump higher than Gulliver.

Part B

Which detail from the article best supports the answer to Part A?

a. "My gentleness and good behaviour had gained so far on the emperor and his court, and indeed upon the army and people in general, that I began to conceive hopes of getting my liberty in a short time."

b. "I was diverted with none so much as that of the rope-dancers, performed upon a slender white thread, extended about two feet, and twelve inches from the ground."

c. "When a great office is vacant, either by death or disgrace (which often happens,) five or six of those candidates petition the emperor to entertain his majesty and the court with a dance on the rope; and whoever jumps the highest, without falling, succeeds in the office."

d. "They are trained in this art from their youth, and are not always of noble birth, or liberal education."

Questions 7-12 are based upon the following passage:

This excerpt is adaptation of Robert Louis Stevenson's *The Strange Case of Dr. Jekyll and Mr. Hyde.*

"Did you ever come across a protégé of his—one Hyde?" He asked.

"Hyde?" repeated Lanyon. "No. Never heard of him. Since my time."

That was the amount of information that the lawyer carried back with him to the great, dark bed on which he tossed to and fro until the small hours of the morning began to grow large. It was a night of little ease to his toiling mind, toiling in mere darkness and besieged by questions.

Six o'clock struck on the bells of the church that was so conveniently near to Mr. Utterson's dwelling, and still he was digging at the problem. Hitherto it had touched him on the intellectual side alone; but; but now his imagination also was engaged, or rather enslaved; and as he lay and tossed in the gross darkness of the night in the curtained room, Mr. Enfield's tale went by before his mind in a scroll of lighted pictures. He would be aware of the great field of lamps in a nocturnal city; then of the figure of a man walking swiftly; then of a child running from the doctor's; and then these met, and that human Juggernaut trod the child down and passed on regardless of her screams. Or else he would see a room in a rich house, where his friend lay asleep, dreaming and smiling at his dreams; and then the door of that room would be opened, the curtains of the bed plucked apart, the sleeper recalled, and, lo! There would stand by his side a figure to whom power was given, and even at that dead hour he must rise and do its bidding. The figure in these two phrases haunted the lawyer all night; and if at anytime he dozed over, it was but to see it glide more stealthily through sleeping houses, or move the more swiftly, and still the more smoothly, even to dizziness, through wider labyrinths of lamplighted city, and at every street corner crush a child and leave her screaming. And still the figure had no face by which he might know it; even in his dreams it had no face, or one that baffled him and melted before his eyes; and thus there it was that there sprung up and grew apace in the lawyer's mind a singularly strong, almost an inordinate, curiosity to behold the features of the real Mr. Hyde. If he could but once set eyes on him, he thought the mystery would lighten and perhaps roll altogether away, as was the habit of mysterious things when well examined. He might see a reason for his friend's strange preference or bondage, and even for the startling clauses of the will. And at least it would be a face worth seeing: the face of a man who was without bowels of mercy: a face which had but to show itself to raise up, in the mind of the unimpressionable Enfield, a spirit of enduring hatred.

From that time forward, Mr. Utterson began to haunt the door in the by street of shops. In the morning before office hours, at noon when business was plenty of time scarce, at night under the face of the full city moon, by all lights and at all hours of solitude or concourse, the lawyer was to be found on his chosen post.

"If he be Mr. Hyde," he had thought, "I should be Mr. Seek."

7. What is the purpose of the use of repetition in the following passage?
    It was a night of little ease to his toiling mind, toiling in mere darkness and besieged by questions.

    a. It serves as a demonstration of the mental state of Mr. Lanyon.
    b. It is reminiscent of the church bells that are mentioned in the story.
    c. It mimics Mr. Utterson's ambivalence.
    d. It emphasizes Mr. Utterson's anguish in failing to identify Hyde's whereabouts.

8.
Part A
What is the setting of the story in this passage?
    a. In the city
    b. On the countryside
    c. In a jail
    d. In a mental health facility

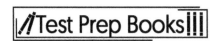

Part B

Which detail from the article best supports the answer to Part A?
- a. "It was a night of little ease to his toiling mind"
- b. "He would be aware of the great field of lamps in a nocturnal city"
- c. "Mr. Enfield's tale went by before his mind in a scroll of lighted pictures."
- d. "And still the figure had no face by which he might know it"

9. What can one infer about the meaning of the word "Juggernaut" from the author's use of it in the passage?
- a. It is an apparition that appears at daybreak.
- b. It scares children.
- c. It is associated with space travel.
- d. Mr. Utterson finds it soothing.

10. What is the definition of the word *haunt* in the following passage?

> From that time forward, Mr. Utterson began to haunt the door in the by street of shops. In the morning before office hours, at noon when business was plenty of time scarce, at night under the face of the full city moon, by all lights and at all hours of solitude or concourse, the lawyer was to be found on his chosen post.

- a. To levitate
- b. To constantly visit
- c. To terrorize
- d. To daunt

11. The phrase *labyrinths of lamplighted city* contains an example of what?
- a. Hyperbole
- b. Simile
- c. Juxtaposition
- d. Alliteration

12. What can one reasonably conclude from the final comment of this passage?

> "If he be Mr. Hyde," he had thought, "I should be Mr. Seek."

- a. The speaker is considering a name change.
- b. The speaker is experiencing an identity crisis.
- c. The speaker has mistakenly been looking for the wrong person.
- d. The speaker intends to continue to look for Hyde.

Questions 13-18 are based upon the following passage:

This excerpt is adaptation from "What to the Slave is the Fourth of July?" Rochester, New York July 5, 1852

> Fellow citizens—Pardon me, and allow me to ask, why am I called upon to speak here today? What have I, or those I represent, to do with your national independence? Are the great principles of political freedom and of natural justice embodied in that Declaration of Independence, Independence extended to us? And am I therefore called upon to bring our humble offering to the national altar, and to confess the benefits, and express devout gratitude for the blessings, resulting from your independence to us?

Would to God, both for your sakes and ours, ours that an affirmative answer could be truthfully returned to these questions! Then would my task be light, and my burden easy and delightful. For who is there so cold that a nation's sympathy could not warm him? Who so obdurate and dead to the claims of gratitude that would not thankfully acknowledge such priceless benefits? Who so stolid and selfish, that would not give his voice to swell the hallelujahs of a nation's jubilee, when the chains of servitude had been torn from his limbs? I am not that man. In a case like that, the dumb may eloquently speak, and the lame man leap as an hart.

But, such is not the state of the case. I say it with a sad sense of the disparity between us. I am not included within the pale of this glorious anniversary. Oh pity! Your high independence only reveals the immeasurable distance between us. The blessings in which you this day rejoice, I do not enjoy in common. The rich inheritance of justice, liberty, prosperity, and independence, bequeathed by your fathers, is shared by *you*, not by *me*. This Fourth of July is *yours*, not *mine*. You may rejoice, *I* must mourn. To drag a man in fetters into the grand illuminated temple of liberty, and call upon him to join you in joyous anthems, were inhuman mockery and sacrilegious irony. Do you mean, citizens, to mock me, by asking me to speak today? If so there is a parallel to your conduct. And let me warn you that it is dangerous to copy the example of a nation whose crimes, towering up to heaven, were thrown down by the breath of the Almighty, burying that nation and irrecoverable ruin! I can today take up the plaintive lament of a peeled and woe-smitten people.

By the rivers of Babylon, there we sat down. Yea! We wept when we remembered Zion. We hanged our harps upon the willows in the midst thereof. For there, they that carried us away captive, required of us a song; and they who wasted us required of us mirth, saying, "Sing us one of the songs of Zion." How can we sing the Lord's song in a strange land? If I forget thee, O Jerusalem, let my right hand forget her cunning. If I do not remember thee, let my tongue cleave to the roof of my mouth.

13.
Part A
What is the tone of the first paragraph of this passage?
   a. Exasperated
   b. Inclusive
   c. Contemplative
   d. Nonchalant

Part B
Which detail from the article best supports the answer to Part A?
   a. "We hanged our harps upon the willows in the midst thereof."
   b. "For there, they that carried us away captive, required of us a song"
   c. "By the rivers of Babylon, there we sat down."
   d. "What have I, or those I represent, to do with your national independence?"

14. Which word CANNOT be used synonymously with the term *obdurate* as it is conveyed in the text below?

> Who so obdurate and dead to the claims of gratitude, that would not thankfully acknowledge such priceless benefits?

    a. Steadfast
    b. Stubborn
    c. Contented
    d. Unwavering

15.
Part A
What is the central purpose of this text?
    a. To demonstrate the author's extensive knowledge of the Bible
    b. To address the feelings of exclusion expressed by African Americans after the establishment of the Fourth of July holiday
    c. To convince wealthy landowners to adopt new holiday rituals
    d. To explain why minorities often relished the notion of segregation in government institutions

Part B
Which detail from the article best supports the answer to Part A?
    a. "Would to God, both for your sakes and ours, ours that an affirmative answer could be truthfully returned to these questions!"
    b. "For who is there so cold that a nation's sympathy could not warm him?"
    c. "The rich inheritance of justice, liberty, prosperity, and independence, bequeathed by your fathers, is shared by *you*, not by *me*."
    d. "If I do not remember thee, let my tongue cleave to the roof of my mouth."

16.
Part A
How does the author provide effective support for the central idea of the passage?
    a. The author offers gratitude for the crowd asking him to speak.
    b. The author examines the reasons why he has gone to Babylon.
    c. The author admits that he is not the right man to speak at the celebration.
    d. The author states that it is insulting to have him speak at their gathering.

Part B
Which detail from the article best supports the answer to Part A?
    a. "Are the great principles of political freedom and of natural justice embodied in that Declaration of Independence, Independence extended to us?"
    b. "Would to God, both for your sakes and ours, ours that an affirmative answer could be truthfully returned to these questions!"
    c. "To drag a man in fetters into the grand illuminated temple of liberty, and call upon him to join you in joyous anthems, were inhuman mockery and sacrilegious irony"
    d. "How can we sing the Lord's song in a strange land?"

17. The statement below features an example of which of the following literary devices?
Oh pity! Your high independence only reveals the immeasurable distance between us.

 a. Assonance
 b. Parallelism
 c. Amplification
 d. Hyperbole

18.
Part A
The speaker's use of biblical references, such as "rivers of Babylon" and the "songs of Zion," helps the reader to do all of the following EXCEPT:
 a. Identify with the speaker through the use of common text.
 b. Convince the audience that injustices have been committed by referencing another group of people who have been previously affected by slavery.
 c. Display the equivocation of the speaker and those that he represents.
 d. Appeal to the listener's sense of humanity.

Part B
Which detail from the article best supports the answer to Part A?
 a. "Would to God, both for your sakes and ours, ours that an affirmative answer could be truthfully returned to these questions!"
 b. "For who is there so cold that a nation's sympathy could not warm him?"
 c. "The rich inheritance of justice, liberty, prosperity, and independence, bequeathed by your fathers, is shared by *you*, not by *me*."
 d. "If I do not remember thee, let my tongue cleave to the roof of my mouth."

Questions 19-24 are based upon the following passage:

This excerpt is an adaptation from Abraham Lincoln's Address Delivered at the Dedication of the Cemetery at Gettysburg, November 19, 1863.

> Four score and seven years ago our fathers brought forth on this continent, a new nation, conceived in liberty, and dedicated to the proposition that all men are created equal.
>
> Now we are engaged in a great civil war, testing whether that nation, or any nation so conceived and so dedicated, can long endure. We are met on a great battlefield of that war. We have come to dedicate a portion of that field, as a final resting place for those who here gave their lives that this nation might live. It is altogether fitting and proper that we should do this.
>
> But, in a larger sense, we cannot dedicate—we cannot consecrate that we cannot hallow—this ground. The brave men, living and dead, who struggled here, have consecrated it, far above our poor power to add or detract. The world will little note, nor long remember what we say here, but it can never forget what they did here. It is for us the living, rather, to be dedicated here to the unfinished work which they who fought here have thus far so nobly advanced. It is rather for us to be here and dedicated to the great task remaining before us—that from these honored dead we take increased devotion to that cause for which they gave the last full measure of devotion—that we

here highly resolve that these dead shall not have died in vain—that these this nation, under God, shall have a new birth of freedom—and that government of people, by the people, for the people, shall not perish from the earth.

19. The best description for the phrase *four score and seven years ago* is which of the following?
    a. A unit of measurement
    b. A period of time
    c. A literary movement
    d. A statement of political reform

20.
Part A
What is the setting of this text?
    a. A battleship off of the coast of France
    b. A desert plain on the Sahara Desert
    c. A battlefield in North America
    d. The residence of Abraham Lincoln

Part B
Which detail from the article best supports the answer to Part A?
    a. "We are met on a great battlefield of that war."
    b. "It is altogether fitting and proper that we should do this."
    c. "But, in a larger sense, we cannot dedicate—we cannot consecrate that we cannot hallow—this ground."
    d. "The world will little note, nor long remember what we say here"

21. Which war is Abraham Lincoln referring to in the following passage?
    Now we are engaged in a great civil war, testing whether that nation, or any nation so conceived and so dedicated, can long endure.

    a. World War I
    b. The War of the Spanish Succession
    c. World War II
    d. The American Civil War

22.
Part A
What message is the author trying to convey through this address?
    a. The audience should consider the death of the people that fought in the war as an example and perpetuate the ideals of freedom that the soldiers died fighting for.
    b. The audience should honor the dead by establishing an annual memorial service.
    c. The audience should form a militia that would overturn the current political structure.
    d. The audience should forget the lives that were lost and discredit the soldiers.

Part B

Which detail from the article best supports the answer to Part A?

a. "We have come to dedicate a portion of that field, as a final resting place for those who here gave their lives that this nation might live."

b. "The brave men, living and dead, who struggled here, have consecrated it, far above our poor power to add or detract."

c. "that government of people, by the people, for the people, shall not perish from the earth."

d. "that from these honored dead we take increased devotion to that cause for which they gave the last full measure of devotion"

23. Which rhetorical device is being used in the following passage?

> . . . we here highly resolve that these dead shall not have died in vain—that these this nation, under God, shall have a new birth of freedom—and that government of people, by the people, for the people, shall not perish from the earth.

a. Antimetabole
b. Antiphrasis
c. Anaphora
d. Epiphora

24. What is the effect of Lincoln's statement in the following passage?

> But, in a larger sense, we cannot dedicate—we cannot consecrate that we cannot hallow—this ground. The brave men, living and dead, who struggled here, have consecrated it, far above our poor power to add or detract.

a. His comparison emphasizes the great sacrifice of the soldiers who fought in the war.
b. His comparison serves as a reminder of the inadequacies of his audience.
c. His comparison serves as a catalyst for guilt and shame among audience members.
d. His comparison attempts to illuminate the great differences between soldiers and civilians.

Questions 25-30 are based upon the following passage:

This excerpt is adaptation from Charles Dickens' speech in Birmingham in England on December 30, 1853 on behalf of the Birmingham and Midland Institute.

> My Good Friends,—When I first imparted to the committee of the projected Institute my particular wish that on one of the evenings of my readings here the main body of my audience should be composed of working men and their families, I was animated by two desires; first, by the wish to have the great pleasure of meeting you face to face at this Christmas time, and accompany you myself through one of my little Christmas books; and second, by the wish to have an opportunity of stating publicly in your presence, and in the presence of the committee, my earnest hope that the Institute will, from the beginning, recognise one great principle—strong in reason and justice—which I believe to be essential to the very life of such an Institution. It is, that the working man shall, from the first unto the last, have a share in the management of an Institution which is designed for his benefit, and which calls itself by his name.
>
> I have no fear here of being misunderstood—of being supposed to mean too much in this. If there ever was a time when any one class could of itself do much for its own good, and for the welfare of society—which I greatly doubt—that time is

44

unquestionably past. It is in the fusion of different classes, without confusion; in the bringing together of employers and employed; in the creating of a better common understanding among those whose interests are identical, who depend upon each other, who are vitally essential to each other, and who never can be in unnatural antagonism without deplorable results, that one of the chief principles of a Mechanics' Institution should consist. In this world a great deal of the bitterness among us arises from an imperfect understanding of one another. Erect in Birmingham a great Educational Institution, properly educational; educational of the feelings as well as of the reason; to which all orders of Birmingham men contribute; in which all orders of Birmingham men meet; wherein all orders of Birmingham men are faithfully represented—and you will erect a Temple of Concord here which will be a model edifice to the whole of England.

Contemplating as I do the existence of the Artisans' Committee, which not long ago considered the establishment of the Institute so sensibly, and supported it so heartily, I earnestly entreat the gentlemen—earnest I know in the good work, and who are now among us,—by all means to avoid the great shortcoming of similar institutions; and in asking the working man for his confidence, to set him the great example and give him theirs in return. You will judge for yourselves if I promise too much for the working man, when I say that he will stand by such an enterprise with the utmost of his patience, his perseverance, sense, and support; that I am sure he will need no charitable aid or condescending patronage; but will readily and cheerfully pay for the advantages which it confers; that he will prepare himself in individual cases where he feels that the adverse circumstances around him have rendered it necessary; in a word, that he will feel his responsibility like an honest man, and will most honestly and manfully discharge it. I now proceed to the pleasant task to which I assure you I have looked forward for a long time.

25. Which word is most closely synonymous with the word *patronage* as it appears in the following statement?

> . . . that I am sure he will need no charitable aid or condescending patronage

a. Auspices
b. Aberration
c . Acerbic
d. Adulation

26.
Part A
Which term is most closely aligned with the definition of the term *working man* as it is defined in the following passage?

> You will judge for yourselves if I promise too much for the working man, when I say that he will stand by such an enterprise with the utmost of his patience, his perseverance, sense, and support . . .

a. Plebeian
b. Viscount
c. Entrepreneur
d. Bourgeois

Part B

Which of the following details best supports the answer to Part A?

a. A working man is not someone who works for institutions or corporations, but someone who is well versed in the workings of the soul.

b. A working man is someone who is probably not involved in social activities because the physical demand for work is too high.

c. A working man is someone who works for wages among the middle class.

d. The working man has historically taken to the field, to the factory, and now to the screen.

27.

Part A

What does the author propose to the audience to do in paragraph 2?

a. The author wishes to read another story from his collection that has to do with economic strife and discord.

b. The author proposes to establish a new committee that will represent the working man in all their affairs.

c. The author proposes the audience build something that will bring together the working man and the employers.

d. The author wishes for the audience to build more schools for children so that they will have a better chance at getting work when they are older.

Part B

Which of the following details best supports the answer to Part A?

a. "and you will erect a Temple of Concord here which will be a model edifice to the whole of England."

b. "I have no fear here of being misunderstood"

c. "In this world a great deal of the bitterness among us arises from an imperfect understanding of one another."

d. "Erect in Birmingham a great Educational Institution, properly educational"

28. Based upon the contextual evidence provided in the passage above, what is the meaning of the term *enterprise* in the third paragraph?

a. Company
b. Courage
c. Game
d. Cause

29. The speaker addresses his audience as *My Good Friends*—what kind of credibility does this salutation give to the speaker?

a. The speaker is an employer addressing his employees, so the salutation is a way for the boss to bridge the gap between himself and his employees.

b. The speaker's salutation is one from an entertainer to his audience and uses the friendly language to connect to his audience before a serious speech.

c. The salutation gives the serious speech that follows a somber tone, as it is used ironically.

d. The speech is one from a politician to the public, so the salutation is used to grab the audience's attention.

30. According to the aforementioned passage, what is the speaker's second desire for his time in front of the audience?
    a. To read a Christmas story
    b. For the working man to have a say in his institution which is designed for his benefit
    c. To have an opportunity to stand in their presence
    d. For the life of the institution to be essential to the audience as a whole

Questions 31-36 are based upon the following passage:

This excerpt is adaptation from *Our Vanishing Wildlife*, by William T. Hornaday

Three years ago, I think there were not many bird-lovers in the United States, who believed it possible to prevent the total extinction of both egrets from our fauna. All the known rookeries accessible to plume-hunters had been totally destroyed. Two years ago, the secret discovery of several small, hidden colonies prompted William Dutcher, President of the National Association of Audubon Societies, and Mr. T. Gilbert Pearson, Secretary, to attempt the protection of those colonies. With a fund contributed for the purpose, wardens were hired and duly commissioned. As previously stated, one of those wardens was shot dead in cold blood by a plume hunter. The task of guarding swamp rookeries from the attacks of money-hungry desperadoes to whom the accursed plumes were worth their weight in gold, is a very chancy proceeding. There is now one warden in Florida who says that "before they get my rookery they will first have to get me."

Thus far the protective work of the Audubon Association has been successful. Now there are twenty colonies, which contain all told, about 5,000 egrets and about 120,000 herons and ibises which are guarded by the Audubon wardens. One of the most important is on Bird Island, a mile out in Orange Lake, central Florida, and it is ably defended by Oscar E. Baynard. To-day, the plume hunters who do not dare to raid the guarded rookeries are trying to study out the lines of flight of the birds, to and from their feeding-grounds, and shoot them in transit. Their motto is—"Anything to beat the law, and get the plumes." It is there that the state of Florida should take part in the war.

The success of this campaign is attested by the fact that last year a number of egrets were seen in eastern Massachusetts—for the first time in many years. And so to-day the question is, can the wardens continue to hold the plume-hunters at bay?

31. The author's use of first person pronoun in the following text does NOT have which of the following effects?

Three years ago, I think there were not many bird-lovers in the United States, who believed it possible to prevent the total extinction of both egrets from our fauna.

    a. The phrase *I think* acts as a sort of hedging, where the author's tone is less direct and/or absolute.
    b. It allows the reader to more easily connect with the author.
    c. It encourages the reader to empathize with the egrets.
    d. It distances the reader from the text by overemphasizing the story.

32.
Part A
What purpose does the quote serve at the end of the first paragraph?

a. The quote shows proof of a hunter threatening one of the wardens.
b. The quote lightens the mood by illustrating the colloquial language of the region.
c. The quote provides an example of a warden protecting one of the colonies.
d. The quote provides much needed comic relief in the form of a joke.

Part B
Which detail from the article best supports the answer to Part A?

a. "It is there that the state of Florida should take part in the war."
b. "The task of guarding swamp rookeries from the attacks of money-hungry desperadoes"
c. "one of those wardens was shot dead in cold blood by a plume hunter."
d. "Thus far the protective work of the Audubon Association has been successful."

33. What is the meaning of the word *rookeries* in the following text?
To-day, the plume hunters who do not dare to raid the guarded rookeries are trying to study out the lines of flight of the birds, to and from their feeding-grounds, and shoot them in transit.

a. Houses in a slum area
b. A place where hunters gather to trade tools
c. A place where wardens go to trade stories
d. A colony of breeding birds

34.
Part A
What is on Bird Island?

a. Hunters selling plumes
b. An important bird colony
c. Bird Island Battle between the hunters and the wardens
d. An important egret with unique plumes

Part B
Which detail from the article best supports the answer to Part A?

a. "Three years ago, I think there were not many bird-lovers in the United States, who believed it possible to prevent the total extinction of both egrets from our fauna."
b. "With a fund contributed for the purpose, wardens were hired and duly commissioned."
c. "The success of this campaign is attested by the fact that last year a number of egrets were seen in eastern Massachusetts"
d. "Now there are twenty colonies, which contain all told, about 5,000 egrets and about 120,000 herons and ibises which are guarded by the Audubon wardens. One of the most important is on Bird Island"

35.
Part A
What is the main purpose of the passage?
    a. To persuade the audience to act in preservation of the bird colonies
    b. To show the effect hunting egrets has had on the environment
    c. To argue that the preservation of bird colonies has had a negative impact on the environment.
    d. To demonstrate the success of the protective work of the Audubon Association

Part B
Which detail from the article best supports the answer to Part A?
    a. "Three years ago, I think there were not many bird-lovers in the United States, who believed it possible to prevent the total extinction of both egrets from our fauna."
    b. "With a fund contributed for the purpose, wardens were hired and duly commissioned."
    c. "The success of this campaign is attested by the fact that last year a number of egrets were seen in eastern Massachusetts—for the first time in many years."
    d. "And so to-day the question is, can the wardens continue to hold the plume-hunters at bay?"

36.
Part A
Why are hunters trying to study the lines of flight of the birds?
    a. To study ornithology, one must know the lines of flight that birds take.
    b. To help wardens preserve the lives of the birds
    c. To have a better opportunity to hunt the birds
    d. To builds their homes under the lines of flight because they believe it brings good luck

Part B
Which detail from the article best supports the answer to Part A?
    a. "All the known rookeries accessible to plume-hunters had been totally destroyed."
    b. "The task of guarding swamp rookeries from the attacks of money-hungry desperadoes to whom the accursed plumes were worth their weight in gold, is a very chancy proceeding."
    c. "There is now one warden in Florida who says that 'before they get my rookery they will first have to get me.'"
    d. "To-day, the plume hunters who do not dare to raid the guarded rookeries are trying to study out the lines of flight of the birds, to and from their feeding-grounds, and shoot them in transit."

Questions 37-42 are based upon the following passage:

This excerpt is adaptation from *The Life-Story of Insects,* by Geo H. Carpenter.

> Insects as a whole are preeminently creatures of the land and the air. This is shown not only by the possession of wings by a vast majority of the class, but by the mode of breathing to which reference has already been made, a system of branching air-tubes carrying atmospheric air with its combustion-supporting oxygen to all the insect's tissues. The air gains access to these tubes through a number of paired air-holes or spiracles, arranged segmentally in series.

> It is of great interest to find that, nevertheless, a number of insects spend much of their time under water. This is true of not a few in the perfect winged state, as for example aquatic beetles and water-bugs ('boatmen' and 'scorpions') which have some way of

protecting their spiracles when submerged, and, possessing usually the power of flight, can pass on occasion from pond or stream to upper air. But it is advisable in connection with our present subject to dwell especially on some insects that remain continually under water till they are ready to undergo their final moult and attain the winged state, which they pass entirely in the air. The preparatory instars of such insects are aquatic; the adult instar is aerial. All may-flies, dragon-flies, and caddis-flies, many beetles and two-winged flies, and a few moths thus divide their life-story between the water and the air. For the present we confine attention to the Stone-flies, the May-flies, and the Dragon-flies, three well-known orders of insects respectively called by systematists the Plecoptera, the Ephemeroptera and the Odonata.

In the case of many insects that have aquatic larvae, the latter are provided with some arrangement for enabling them to reach atmospheric air through the surface-film of the water. But the larva of a stone-fly, a dragon-fly, or a may-fly is adapted more completely than these for aquatic life; it can, by means of gills of some kind, breathe the air dissolved in water.

37.
Part A
Which statement best details the central idea in this passage?
    a. It introduces certain insects that transition from water to air.
    b. It delves into entomology, especially where gills are concerned.
    c. It defines what constitutes as insects' breathing.
    d. It invites readers to have a hand in the preservation of insects.

Part B
Which detail from the article best supports the answer to Part A?
    a. "Insects as a whole are preeminently creatures of the land and the air."

    b. "For the present we confine attention to the Stone-flies, the May-flies, and the Dragon-flies, three well-known orders of insects respectively called by systematists the Plecoptera, the Ephemeroptera and the Odonata."
    c. "All may-flies, dragon-flies, and caddis-flies, many beetles and two-winged flies, and a few moths thus divide their life-story between the water and the air."
    d. "In the case of many insects that have aquatic larvae, the latter are provided with some arrangement for enabling them to reach atmospheric air through the surface-film of the water."

38. Which definition most closely relates to the usage of the word *moult* in the passage?
    a. An adventure of sorts, especially underwater
    b. Mating act between two insects
    c. The act of shedding part or all of the outer shell
    d. Death of an organism that ends in a revival of life

39.
Part A
What is the purpose of the first paragraph in relation to the second paragraph?

a. The first paragraph serves as a cause and the second paragraph serves as an effect.

b. The first paragraph serves as a contrast to the second.

c. The first paragraph is a description for the argument in the second paragraph.

d. The first and second paragraphs are merely presented in a sequence.

Part B
Which detail from the article best supports the answer to Part A?

a. "It is of great interest to find that, nevertheless, a number of insects spend much of their time under water."

b. "The air gains access to these tubes through a number of paired air-holes or spiracles, arranged segmentally in series."

c. "The preparatory instars of such insects are aquatic; the adult instar is aerial."

d. "In the case of many insects that have aquatic larvae, the latter are provided with some arrangement for enabling them to reach atmospheric air through the surface-film of the water."

40. What does the following sentence most nearly mean?

The preparatory instars of such insects are aquatic; the adult instar is aerial.

a. The volume of water is necessary to prep the insect for transition rather than the volume of the air.

b. The abdomen of the insect is designed like a star in the water as well as the air.

c. The stage of preparation in between molting is acted out in the water, while the last stage is in the air.

d. These insects breathe first in the water through gills, yet continue to use the same organs to breathe in the air.

41. Which of the statements reflect information that one could reasonably infer based on the author's tone?

a. The author's tone is persuasive and attempts to call the audience to action.

b. The author's tone is passionate due to excitement over the subject and personal narrative.

c. The author's tone is informative and exhibits interest in the subject of the study.

d. The author's tone is somber, depicting some anger at the state of insect larvae.

42. Which statement best describes stoneflies, mayflies, and dragonflies?

a. They are creatures of the land and the air.

b. They have a way of protecting their spiracles when submerged.

c. Their larvae can breathe the air dissolved in water through gills of some kind.

d. The preparatory instars of these insects are aerial.

Questions 43-47 are based upon the following passage:

This excerpt is adaptation from "The 'Hatchery' of the Sun-Fish"--- *Scientific American*, #711

I have thought that an example of the intelligence (instinct?) of a class of fish which has come under my observation during my excursions into the Adirondack region of New

York State might possibly be of interest to your readers, especially as I am not aware that any one except myself has noticed it, or, at least, has given it publicity.

The female sun-fish (called, I believe, in England, the roach or bream) makes a "hatchery" for her eggs in this wise. Selecting a spot near the banks of the numerous lakes in which this region abounds, and where the water is about 4 inches deep, and still, she builds, with her tail and snout, a circular embankment 3 inches in height and 2 thick. The circle, which is as perfect a one as could be formed with mathematical instruments, is usually a foot and a half in diameter; and at one side of this circular wall an opening is left by the fish of just sufficient width to admit her body.

The mother sun-fish, having now built or provided her "hatchery," deposits her spawn within the circular inclosure, and mounts guard at the entrance until the fry are hatched out and are sufficiently large to take charge of themselves. As the embankment, moreover, is built up to the surface of the water, no enemy can very easily obtain an entrance within the inclosure from the top; while there being only one entrance, the fish is able, with comparative ease, to keep out all intruders.

I have, as I say, noticed this beautiful instinct of the sun-fish for the perpetuity of her species more particularly in the lakes of this region; but doubtless the same habit is common to these fish in other waters.

43.
Part A
What is the purpose of this passage?
    a. To show the effects of fish hatcheries on the Adirondack region
    b. To persuade the audience to study Ichthyology (fish science)
    c. To depict the sequence of mating among sun-fish
    d. To enlighten the audience on the habits of sun-fish and their hatcheries

Part B
Which detail from the article best supports the answer to Part A?
    a. "Selecting a spot near the banks of the numerous lakes in which this region abounds, and where the water is about 4 inches deep, and still, she builds, with her tail and snout, a circular embankment 3 inches in height and 2 thick."
    b. "The circle, which is as perfect a one as could be formed with mathematical instruments, is usually a foot and a half in diameter; and at one side of this circular wall an opening is left by the fish of just sufficient width to admit her body."
    c. "I have thought that an example of the intelligence (instinct?) of a class of fish which has come under my observation during my excursions into the Adirondack region of New York State might possibly be of interest to your readers"
    d. "As the embankment, moreover, is built up to the surface of the water, no enemy can very easily obtain an entrance within the inclosure from the top"

44. What does the word *wise* in this passage most closely mean?
    a. Knowledge
    b. Manner
    c. Shrewd
    d. Ignorance

45. What is the definition of the word *fry* as it appears in the following passage?

> The mother sun-fish, having now built or provided her "hatchery," deposits her spawn within the circular inclosure, and mounts guard at the entrance until the fry are hatched out and are sufficiently large to take charge of themselves.

a. Fish at the stage of development where they are capable of feeding themselves.
b. Fish eggs that have been fertilized.
c. A place where larvae is kept out of danger from other predators.
d. A dish where fish is placed in oil and fried until golden brown.

46.

Part A

How is the circle that keeps the larvae of the sun-fish made?

a. It is formed with mathematical instruments.
b. The sun-fish builds it with her tail and snout.
c. It is provided to her as a "hatchery" by Mother Nature.
d. The sun-fish builds it with her larvae.

Part B

Which detail from the article best supports the answer to Part A?

a. "at one side of this circular wall an opening is left by the fish of just sufficient width to admit her body."
b. "The mother sun-fish, having now built or provided her "hatchery," deposits her spawn within the circular inclosure"
c. "she builds, with her tail and snout, a circular embankment 3 inches in height and 2 thick."
d. "while there being only one entrance, the fish is able, with comparative ease, to keep out all intruders."

47. The author included the third paragraph in the following passage to achieve which of the following effects?

a. To complicate the subject matter
b. To express a bias
c. To insert a counterargument
d. To conclude a sequence and add a final detail

Read this article about NASA technology and answer questions 48 – 50.

When researchers and engineers undertake a large-scale scientific project, they may end up making discoveries and developing technologies that have far wider uses than originally intended. This is especially true in NASA, one of the most influential and innovative scientific organizations in America. NASA *spinoff technology* refers to innovations originally developed for NASA space projects that are now used in a wide range of different commercial fields. Many consumers are unaware that products they are buying are based on NASA research!

The first spinoff technology worth mentioning is baby food. In space, where astronauts have limited access to fresh food and fewer options about their daily meals, malnutrition is a serious concern. Consequently, NASA researchers were looking for ways to enhance the nutritional value of astronauts' food. Scientists found that a certain type of algae could be added to food, improving the food's neurological benefits. When experts in the commercial food industry learned of this algae's potential to boost brain health, they were quick to begin their own research. The nutritional substance from algae then developed into a product called life's DHA, which can be found in over 90 percent of infant food sold in America.

Another intriguing example of a spinoff technology can be found in fashion. People who are always dropping their sunglasses may have invested in a pair of sunglasses with scratch resistant lenses—that is, it's impossible to scratch the glass, even if the glasses are dropped on an abrasive surface. This innovation is incredibly advantageous for people who are clumsy, but most shoppers don't know that this technology was originally developed by NASA. Scientists first created scratch resistant glass to help protect costly and crucial equipment from getting scratched in space, especially the helmet visors in space suits. However, sunglasses companies later realized that this technology could be profitable for their products, and they licensed the technology from NASA.

48.
Part A
What is the main purpose of this article?
    a. To advise consumers to do more research before making a purchase
    b. To persuade readers to support NASA research
    c. To tell a narrative about the history of space technology
    d. To define and describe examples of spinoff technology

Part B
Which detail from the article best supports the answer to Part A?
    a. "The first spinoff technology worth mentioning is baby food."

    b. "NASA *spinoff technology* refers to innovations originally developed for NASA space projects that are now used in a wide range of different commercial fields."
    c. "The nutritional substance from algae then developed into a product called life's DHA, which can be found in over 90 percent of infant food sold in America."
    d. "Scientists first created scratch resistant glass to help protect costly and crucial equipment from getting scratched in space"

49. What is the organizational structure of this article?
    a. A general definition followed by more specific examples
    b. A general opinion followed by supporting arguments
    c. An important moment in history followed by chronological details
    d. A popular misconception followed by counterevidence

50.
Part A
Why did NASA scientists research algae?
    a. They already knew algae was healthy for babies.
    b. They were interested in how to grow food in space.
    c. They were looking for ways to add health benefits to food.
    d. They hoped to use it to protect expensive research equipment.

Part B

Which detail from the article best supports the answer to Part A?

a. "People who are always dropping their sunglasses may have invested in a pair of sunglasses with scratch resistant lenses"

b. "When experts in the commercial food industry learned of this algae's potential to boost brain health, they were quick to begin their own research."

c. "This innovation is incredibly advantageous for people who are clumsy, but most shoppers don't know that this technology was originally developed by NASA."

d. "Scientists found that a certain type of algae could be added to food, improving the food's neurological benefits."

Answer Explanations

1. Part A: C: Gulliver becomes acquainted with the people and practices of his new surroundings. Choice *C* is the correct answer because it most extensively summarizes the entire passage. While Choices *A* and *B* are reasonable possibilities, they reference portions of Gulliver's experiences, not the whole. Choice *D* is incorrect because Gulliver doesn't express repentance or sorrow in this particular passage.

Part B: D: Gulliver mentions in the very first sentence of the passage his intentions of "becoming acquainted with the people and practices of his new surroundings." Many times, the central idea of the passage will be located in the first sentence, although this is not always the case. The rest of the choices describe a specific detail in the text and are not part of a central idea.

2. A: Principal refers to *chief* or *primary* within the context of this text. Choice *A* is the answer that most closely aligns with this answer. Choices *B* and *D* make reference to a helper or followers while Choice *C* doesn't meet the description of Gulliver from the passage.

3. Part A: C: One can reasonably infer that Gulliver is considerably larger than the children who were playing around him because multiple children could fit into his hand. Choice *B* is incorrect because there is no indication of stress in Gulliver's tone. Choices *A* and *D* aren't the best answer because though Gulliver seems fond of his new acquaintances, he didn't travel there with the intentions of meeting new people or to express a definite love for them in this particular portion of the text.

Part B: A: From Part A, we see children playing all over Gulliver. We know that what they are doing is friendly because of the answer choice in Part B: Gulliver wished to "cultivate a favourable disposition." The rest of the answer choices are incorrect, as they do not explain how we know the passage in Part A has to do with children playing.

4. C: The emperor made a *definitive decision* to expose Gulliver to their native customs. In this instance, the word *mind* was not related to a vote, question, or cognitive ability.

5. A: Choice *A* is correct. This assertion does *not* support the fact that games are a commonplace event in this culture because it mentions conduct, not games. Choices *B*, *C*, and *D* are incorrect because these do support the fact that games were a commonplace event.

6. Part A: B: Choice *B* is the only option that mentions the correlation between physical ability and leadership positions. Choices *A* and *D* are unrelated to physical strength and leadership abilities. Choice *C* does not make a deduction that would lead to the correct answer—it only comments upon the abilities of common townspeople.

Part B: C: This sentence describes how those successful on the rope are the ones who succeed in office. Part A explains that the two men who were in high offices had skills on the rope, and Choice *C* is the sentence that describes this phenomenon.

7. D: It emphasizes Mr. Utterson's anguish in failing to identify Hyde's whereabouts. Context clues indicate that Choice *D* is correct because the passage provides great detail of Mr. Utterson's feelings about locating Hyde. Choice *A* does not fit because there is no mention of Mr. Lanyon's mental state. Choice *B* is incorrect; although the text does make mention of bells, Choice *B* is not the *best* answer overall. Choice *C* is incorrect because the passage clearly states that Mr. Utterson was determined, not unsure.

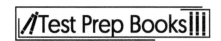

8. Part A: A: In the city. The word *city* appears in the passage several times, thus establishing the location for the reader.

Part B: B: This sentence mentions the word "city." The other answer choices do not mention that the setting takes place in the city, therefore they are incorrect.

9. B: It scares children. The passage states that the Juggernaut causes the children to scream. Choices *A* and *D* don't apply because the text doesn't mention either of these instances specifically. Choice *C* is incorrect because there is nothing in the text that mentions space travel.

10. B: To constantly visit. The mention of *morning, noon,* and *night* make it clear that the word *haunt* refers to frequent appearances at various locations. Choice *A* doesn't work because the text makes no mention of levitating. Choices *C* and *D* are not correct because the text makes mention of Mr. Utterson's anguish and disheartenment because of his failure to find Hyde but does not make mention of Mr. Utterson's feelings negatively affecting anyone else.

11. D: This is an example of alliteration. Choice *D* is the correct answer because of the repetition of the *L*-words. Hyperbole is an exaggeration, so Choice *A* doesn't work. No comparison is being made, so no simile or metaphor is being used, thus eliminating Choices *B* and *C*.

12. D: The speaker intends to continue to look for Hyde. Choices *A* and *B* are not possible answers because the text doesn't refer to any name changes or an identity crisis, despite Mr. Utterson's extreme obsession with finding Hyde. The text also makes no mention of a mistaken identity when referring to Hyde, so Choice *C* is also incorrect.

13. Part A: A: The tone is exasperated. While contemplative is an option because of the inquisitive nature of the text, Choice *A* is correct because the speaker is annoyed by the thought of being included when he felt that the fellow members of his race were being excluded. The speaker is not nonchalant, nor accepting of the circumstances which he describes.

Part B: D: This sentence is a rhetorical question used to show exasperation in the first paragraph of the passage. Choices *A, B,* and *C* are allusions to a biblical reference.

14. C: Choice *C, contented,* is the only word that has different meaning. Furthermore, the speaker expresses objection and disdain throughout the entire text.

15. Part A: B: To address the feelings of exclusion expressed by African Americans after the establishment of the Fourth of July holiday. While the speaker makes biblical references, it is not the main focus of the passage, thus eliminating Choice *A* as an answer. The passage also makes no mention of wealthy landowners and doesn't speak of any positive response to the historical events, so Choices *C* and *D* are not correct.

Part B: C: Choice *C* best supports the main idea, that the speaker addresses feelings of exclusion expressed by African Americans after the establishment of the Fourth of July holiday. Choice *A* is asking the audience to think about the truth. Choice *B* is a rhetorical question asked by the speaker implicating why the speaker does not seem to have gratitude. Choice *D* is a biblical reference used to create pathos.

16. Part A: D: Choice *D* is the correct answer because the author states that to have him speak there is "inhuman mockery" and "sacrilegious irony," which means that it insults him. Choice *A* is incorrect; the author does not offer gratitude to the crowd. Choice *B* is incorrect; the author mentions Babylon at the very end only for the purpose of pathos—it is not a place where he says he has been. Choice *C* is incorrect; the author does admit that he is not the right man to speak for the occasion, but this is not the best answer that provides support for the central idea of the passage, which is to address the feelings of exclusion expressed by African Americans regarding the fourth of July.

Part B: C: Choice C is the best answer because it is the quote that explains *why* African Americans feel exclusion from the fourth of July: "To drag a man in fetters into the grand illuminated temple of liberty, and call upon him to join you in joyous anthems, were inhuman mockery and sacrilegious irony"—this quote explains the mockery and irony felt in front of this celebration. The other answer choices are incorrect.

17. D: Hyperbole. Choices A and B are unrelated. Assonance is the repetition of sounds and commonly occurs in poetry. Parallelism refers to two statements that correlate in some manner. Choice C is incorrect because amplification normally refers to clarification of meaning by broadening the sentence structure, while hyperbole refers to a phrase or statement that is being exaggerated.

18. Part A: C: Display the equivocation of the speaker and those that he represents. Choice C is correct because the speaker is clear about his intention and stance throughout the text. Choice A could be true, but the words "common text" is arguable. Choice B is also partially true, as another group of people affected by slavery are being referenced. However, the speaker is not trying to convince the audience that injustices have been committed, as it is already understood there have been injustices committed. Choice D is also close to the correct answer, but it is not the *best* answer choice possible.

Part B: D: Choice D is the only evidence from the text that has to do with a biblical reference. We can see this by the language used: "thee," which is archaic for this time period. The other answer choices demonstrate the speaker addressing the audience, not drawing upon allusions.

19. B: A period of time. It is apparent that Lincoln is referring to a period of time within the context of the passage because of how the sentence is structured with the word *ago*.

20. Part A: C: Lincoln's reference to *the brave men, living and dead, who struggled here,* proves that he is referring to a battlefield. Choices A and B are incorrect, as a *civil war* is mentioned and not a war with France or a war in the Sahara Desert. Choice D is incorrect because it does not make sense to consecrate a President's ground instead of a battlefield ground for soldiers who died during the American Civil War.

Part B: A: Choice A describes the setting of the passage in question: a battlefield. The term "that war" should also denote that this is a battlefield in North America, since Lincoln is giving a speech over the American Civil War.

21. D: Abraham Lincoln is the former president of the United States, and he references a "civil war" during his address.

22. Part A: A: The audience should consider the death of the people that fought in the war as an example and perpetuate the ideals of freedom that the soldiers died fighting for. Lincoln doesn't address any of the topics outlined in Choices B, C, or D. Therefore, Choice A is the correct answer.

Part B: D: The best answer choice that encompasses considering the death of the soldiers to perpetuate ideas of freedom is Choice D. The other answer choices are close. Choice A says what they have come to do, but it is not the central message that Lincoln wants the audience to take away. Choice B is a statement that is countered in the very next paragraph by the central message. Choice C is just an extra

detail to add to the main idea that concepts of freedom should be perpetuated; it gives a shout out to democracy at the very end.

23. D: Choice *D* is the correct answer because of the repetition of the word *people* at the end of the passage. Choice *A, antimetatabole*, is the repetition of words in a succession. Choice *B, antiphrasis*, is a form of denial of an assertion in a text. Choice *C, anaphora*, is the repetition that occurs at the beginning of sentences.

24. A: Choice *A* is correct because Lincoln's intention was to memorialize the soldiers who had fallen as a result of war as well as celebrate those who had put their lives in danger for the sake of their country. Choices *B* and *D* are incorrect because Lincoln's speech was supposed to foster a sense of pride among the members of the audience while connecting them to the soldiers' experiences.

25. A: The word *patronage* most nearly means *auspices*, which means *protection* or *support*. Choice *B, aberration*, means *deformity* and does not make sense within the context of the sentence. Choice *C, acerbic,* means *bitter* and also does not make sense in the sentence. Choice *D, adulation*, is a positive word meaning *praise*, and thus does not fit with the word *condescending* in the sentence.

26. Part A: D: *Working man* is most closely aligned with Choice *D, bourgeois.* In the context of the speech, the word *bourgeois* means *working* or *middle class*. Choice *A, Plebeian*, does suggest *common people*; however, this is a term that is specific to ancient Rome. Choice *B, viscount*, is a European title used to describe a specific degree of nobility. Choice *C, entrepreneur*, is a person who operates their own business.

Part B: C: In the context of the speech, the term *working man* most closely correlates with Choice *C, working man is someone who works for wages among the middle class.* Choice *A* is not mentioned in the passage and is off-topic. Choice *B* may be true in some cases, but it does not reflect the sentiment described for the term *working man* in the passage. Choice *D* may also be arguably true. However, it is not given as a definition but as *acts* of the working man, and the topics of *field, factory,* and *screen* are not mentioned in the passage.

27. Part A: C: Choice *C* is the correct answer; the author proposes the audience build something that will bring together the working man and the employers. The author's proposal of this is in the last sentence where he mentions the "Temple of Concord," which was a temple built in Rome to bring together the patricians (ruling class) and the plebeians (working class) in a time of economic strife. The author is calling symbolically for this same type of reconciliation in England.

Part B: A: Choice *A* is the correct answer: "and you will erect a Temple of Concord here which will be a model edifice to the whole of England." This quote from the passage best supports the answer to Part A, that the author proposes the audience build something that will bring employees and employers together in harmony.

28. D: *Enterprise* most closely means *cause*. Choices *A, B,* and *C* are all related to the term *enterprise*. However, Dickens speaks of a *cause* here, not a company, courage, or a game. *He will stand by such an enterprise* is a call to stand by a cause to enable the working man to have a certain autonomy over his own economic standing. The very first paragraph ends with the statement that the working man *shall . . . have a share in the management of an institution which is designed for his benefit.*

29. B: The speaker's salutation is one from an entertainer to his audience and uses the friendly language to connect to his audience before a serious speech. Recall in the first paragraph that the speaker is there to "accompany [the audience] . . . through one of my little Christmas books," making him an author there to entertain the crowd with his own writing. The speech preceding the reading is the passage itself, and, as the tone indicates, a serious speech addressing the "working man." Although the passage speaks of employers and employees, the speaker himself is not an employer of the audience, so Choice A is incorrect. Choice C is also incorrect, as the salutation is not used ironically, but sincerely, as the speech addresses the wellbeing of the crowd. Choice D is incorrect because the speech is not given by a politician, but by a writer.

30. B: For the working man to have a say in his institution which is designed for his benefit Choice A is incorrect because that is the speaker's *first* desire, not his second. Choices C and D are tricky because the language of both of these is mentioned after the word *second*. However, the speaker doesn't get to the second wish until the next sentence. Choices C and D are merely prepositions preparing for the statement of the main clause, Choice B.

31. D: The use of "I" could serve to have a "hedging" effect, allow the reader to connect with the author in a more personal way, and cause the reader to empathize more with the egrets. However, it doesn't distance the reader from the text, making Choice D the answer to this question.

32. Part A: C: The quote provides an example of a warden protecting one of the colonies. Choice A is incorrect because the speaker of the quote is a warden, not a hunter. Choice B is incorrect because the quote does not lighten the mood, but shows the danger of the situation between the wardens and the hunters. Choice D is incorrect because there is no humor found in the quote.

Part B: B: Choice B backs up the quote by showing the warden is *guarding* the rookery, and thereby protecting one of the colonies.

33. D: A *rookery* is a colony of breeding birds. Although *rookery* could mean Choice A, houses in a slum area, it does not make sense in this context. Choices B and C are both incorrect, as this is not a place for hunters to trade tools or for wardens to trade stories.

34. Part A: B: An important bird colony. The previous sentence is describing "twenty colonies" of birds, so what follows should be a bird colony. Choice A may be true, but we have no evidence of this in the text. Choice C does touch on the tension between the hunters and wardens, but there is no official "Bird Island Battle" mentioned in the text. Choice D does not exist in the text.

Part B: D: The best detail that supports that there is a bird colony on Bird Island is Choice D: "Now there are twenty colonies, which contain all told, about 5,000 egrets and about 120,000 herons and ibises which are guarded by the Audubon wardens. One of the most important is on Bird Island." The other answer choices do not talk about what's on Bird Island.

35. Part A: D: To demonstrate the success of the protective work of the Audubon Association. The text mentions several different times how and why the association has been successful and gives examples to back this fact. Choice A is incorrect because although the article, in some instances, calls certain people to act, it is not the purpose of the entire passage. There is no way to tell if Choices B and C are correct, as they are not mentioned in the text.

Part B: C: Choice *C* is the correct answer because it is a detail that demonstrates firsthand the success of the protective work of the association. Choices *A, B,* and *D* are incorrect. Choice *A* demonstrates how incredulous the campaign has been. Choice *B* shares a detail of how the association came to be. Choice *D* is a rhetorical question used to make the audience consider the future of the association.

36. Part A: C: To have a better opportunity to hunt the birds. Choice *A* might be true in a general sense, but it is not relevant to the context of the text. Choice *B* is incorrect because the hunters are not studying lines of flight to help wardens, but to hunt birds. Choice *D* is incorrect because nothing in the text mentions that hunters are trying to build homes underneath lines of flight of birds for good luck.

Part B: D: Choice *D* is a detail that shows that the hunters study the flight lines of birds in order to hunt them. The rest of the answer choices are incorrect because they do not state a reason why hunters want to study the flight lines of birds.

37. Part A: A: It introduces certain insects that transition from water to air. Choice *B* is incorrect because although the passage talks about gills, it is not the central idea of the passage. Choices *C* and *D* are incorrect because the passage does not "define" or "invite," but only serves as an introduction to stoneflies, dragonflies, and mayflies and their transition from water to air.

Part B: C: Choice *C* is the only answer choice that names "certain insects that transition from water to air." Notice the "thus," which means that the text preceding this statement has been leading up to it. Although Choice *A* is the first sentence in the passage, it is not evidence of the central idea, because the first paragraph is set up to be a contrast to the main idea. Choices *B* and *D* give more detail about the named insects, but they are not the overall main idea of the passage.

38. C: The act of shedding part or all of the outer shell. Choices *A, B,* and *D* are incorrect.

39. Part A: B: The first paragraph serves as a contrast to the second. Notice how the first paragraph goes into detail describing how insects are able to breathe air. The second paragraph acts as a contrast to the first by stating "[i]t is of great interest to find that, nevertheless, a number of insects spend much of their time under water." Watch for transition words such as "nevertheless" to help find what type of passage you're dealing with.

Part B: A: Choice *A* is correct because we get the word "nevertheless," which indicates a contrast to the text that comes before it. Choice *B* is the last sentence of the first paragraph, but it does not yet establish the contrast with the second paragraph. Choices *C* and *D* are incorrect because they come much later in the passage and therefore do not pertain to Part A of the question.

40: C: The stage of preparation in between molting is acted out in the water, while the last stage is in the air. Choices *A, B,* and *D* are all incorrect. *Instars* is the phase between two periods of molting, and the text explains when these transitions occur.

41. C: The author's tone is informative and exhibits interest in the subject of the study. Overall, the author presents us with information on the subject. One moment where personal interest is depicted is when the author states, "It is of great interest to find that, nevertheless, a number of insects spend much of their time under water.'"

42. C: Their larva can breathe the air dissolved in water through gills of some kind. This is stated in the last paragraph. Choice *A* is incorrect because the text mentions this in a general way at the beginning of the passage concerning "insects as a whole." Choice *B* is incorrect because this is stated of beetles and water-bugs, and not the insects in question. Choice *D* is incorrect because this is the opposite of what the text says of instars.

43. Part A: D: To enlighten the audience on the habits of sun-fish and their hatcheries. Choice *A* is incorrect because although the Adirondack region is mentioned in the text, there is no cause or effect relationships between the region and fish hatcheries depicted here. Choice *B* is incorrect because the text does not have an agenda, but rather is meant to inform the audience. Finally, Choice *C* is incorrect because the text says nothing of how sun-fish mate.

Part B: C: Choice *C* is correct. It is the first sentence of the passage and outlines the purpose of the text: to talk about the habits of a certain fish and how it will be of interest to readers. The other choices listed are only details that support the main idea.

44. B: The word *wise* in this passage most closely means *manner*. Choices *A* and *C* are synonyms of *wise*; however, they are not relevant in the context of the text. Choice *D*, *ignorance*, is opposite of the word *wise*, and is therefore incorrect.

45. A: Fish at the stage of development where they are capable of feeding themselves. Even if the word *fry* isn't immediately known to the reader, the context gives a hint when it says "until the fry are hatched out and are sufficiently large to take charge of themselves."

46. Part A: B: The sun-fish builds it with her tail and snout. The text explains this in the second paragraph. Choice *A* is used in the text as a metaphor. Choices *C* and *D* are incorrect.

Part B: C: Choice *C* is correct; it is a straightforward explanation of how the sun-fish builds the circle. Choice *A* describes the circle after it has been built, but it does not say *how* it is built. Choice *B* describes what she does with the hatchery after she has built it. Choice *D* is another detail of the hatchery that explains a characteristic of it and now *how* it came to be.

47. D: To conclude a sequence and add a final detail. The concluding sequence is expressed in the phrase "[t]he mother sun-fish, having now built or provided her 'hatchery.'" The final detail is the way in which the sun-fish guards the "inclosure." Choices *A, B,* and *C* are incorrect.

48. Part A: D: To define and describe examples of spinoff technology. This is a purpose question—*why* did the author write this? The article contains facts, definitions, and other objective information without telling a story or arguing an opinion. In this case, the purpose of the article is to inform the reader. The only answer choice related to giving information is Choice *D*: to define and describe.

Part B: B: Choice *B* is the correct answer because it defines spinoff technology then lists the products that it will be describing. The other answer choices do list the details of some of the products, but they do not give an overall picture of the purpose of the passage.

49. A: A general definition followed by more specific examples. This organization question asks readers to analyze the structure of the essay. The topic of the essay is spinoff technology; the first paragraph gives a general definition of the concept, while the following two paragraphs offer more detailed examples to help illustrate this idea.

50. Part A: C: They were looking for ways to add health benefits to food. This reading comprehension question can be answered based on the second paragraph—scientists were concerned about astronauts' nutrition and began researching nutritional supplements. Choice *A* isn't true because it reverses the order of discovery (first NASA identified algae for astronaut use, and then it was further developed for use in baby food).

Part B: D: Choice *D* is the correct answer because it explains directly why NASA researched algae: because it could improve neurological benefits. Choice *B* is close, but this is more about how the commercial food industry learned about the benefits of algae. Choices *A* and *C* have to do with another spinoff technology.

Geometry

Know Precise Definitions Relevant to Geometry

Algebraic equations can be used to describe geometric figures in the plane. The method for doing so is to use the *Cartesian coordinate plane*. The idea behind these Cartesian coordinates (named for mathematician and philosopher Descartes) is that from a specific point on the plane, known as the *center*, one can specify any other point by saying *how far to the right or left* and *how far up or down*.

The plane is covered with a grid. The two directions, right to left and bottom to top, are called *axes* (singular *axis*). When working with *x* and *y* variables, the *x* variable corresponds to the right and left axis, and the *y* variable corresponds to the up and down axis.

Any point on the grid is found by specifying how far to travel from the center along the *x*-axis and how far to travel along the *y*-axis. The ordered pair can be written as (x, y). A positive *x* value means go to the right on the *x*-axis, while a negative *x* value means to go to the left. A positive *y* value means to go up, while a negative value means to go down. Several points are shown as examples in the figure.

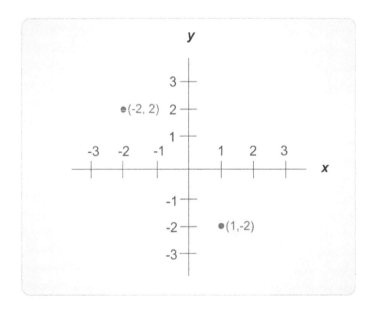

Cartesian Coordinate Plane

The coordinate plane can be divided into four *quadrants*. The upper-right part of the plane is called the *first quadrant*, where both *x* and *y* are positive. The *second quadrant* is the upper-left, where *x* is negative but *y* is positive. The *third quadrant* is the lower left, where both *x* and *y* are negative. Finally,

the *fourth quadrant* is in the lower right, where *x* is positive but *y* is negative. These quadrants are often written with Roman numerals:

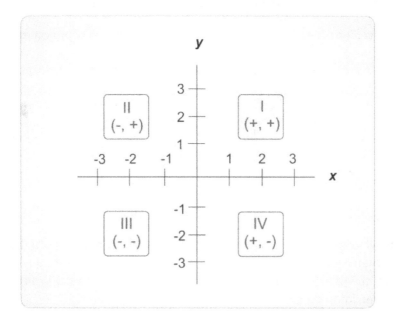

In addition to graphing individual points as shown above, the graph lines and curves in the plane can be graphed corresponding to equations. In general, if there is some equation involving *x* and *y*, then the *graph* of that equation consists of all the points (x, y) in the Cartesian coordinate plane, which satisfy this equation.

Given the equation $y = x + 2$, the point $(0, 2)$ is in the graph, since $2 = 0 + 2$ is a true equation. However, the point $(1, 4)$ will *not* be in the graph, because $4 = 1 + 2$ is false.

Straight Lines
The simplest equations to graph are the equations whose graphs are lines, called *linear equations*. Every linear equation can be rewritten algebraically so that it looks like $Ax + By = C$.

First, the ratio of the change in the *y* coordinate to the change in the *x* coordinate is constant for any two distinct points on the line. In any pair of points on a line, two points, (x_1, y_1) and (x_2, y_2)—

where $x_1 \neq x_2$—the ratio $\frac{y_2 - y_1}{x_2 - x_1}$ will always be the same, even if another pair of points is used.

This ratio, $\frac{y_2 - y_1}{x_2 - x_1}$, is called the *slope* of the line and is often denoted with the letter *m*. If the slope is *positive*, then the line goes upward when moving to the right. If the slope is *negative*, then it moves downward when moving to the right. If the slope is 0, then the line is *horizontal*, and the *y* coordinate is constant along the entire line. For lines where the *x* coordinate is constant along the entire line, the slope is not defined, and these lines are called *vertical* lines.

The *y* coordinate of the point where the line touches the *y*-axis is called the *y-intercept* of the line. It is often denoted by the letter *b*, used in the form of the linear equation $y = mx + b$. The *x* coordinate of the point where the line touches the *x*-axis is called the *x-intercept*. It is also called the *zero* of the line.

Suppose two lines have slopes m_1 and m_2. If the slopes are equal, $m_1 = m_2$, then the lines are *parallel*. Parallel lines never meet one another. If $m_1 = -\frac{1}{m_2}$, then the lines are called *perpendicular* or *orthogonal*. Their slopes can also be called opposite reciprocals of each other.

There are several convenient ways to write down linear equations. The common forms are listed here:

Standard Form: $Ax + By = C$, where the slope is given by $\frac{-A}{B}$, and the *y*-intercept is given by $\frac{C}{B}$.

Slope-Intercept Form: $y = mx + b$, where the slope is *m*, and the *y*-intercept is *b*.

Point-Slope Form: $y - y_1 = m(x - x_1)$, where *m* is the slope, and (x_1, y_1) is any point on the line.

Two-Point Form: $\frac{y - y_1}{x - x_1} = \frac{y_2 - y_1}{x_2 - x_1}$, where (x_1, y_1), and (x_2, y_2) are any two distinct points on the line.

Intercept Form: $\frac{x}{x_1} + \frac{y}{y_1} = 1$, where x_1 is the *x*-intercept, and y_1 is the *y*-intercept.

Depending upon the given information, different forms of the linear equation can be easier to write down than others. When given two points, the two-point form is easy to write down. If the slope and a single point is known, the point-slope form is easiest to start with. In general, which form to start with depends upon the given information.

Conics
The graph of an equation of the form $y = ax^2 + bx + c$ or $x = ay^2 + by + c$ is called a *parabola*.

The graph of an equation of the form $\frac{x^2}{a^2} - \frac{y^2}{b^2} = 1$ or $-\frac{x^2}{a^2} + \frac{y^2}{b^2} = 1$ is called a *hyperbola*.

The graph of an equation of the form $\frac{(x - x_0)^2}{a^2} + \frac{(y - y_0)^2}{b^2} = 1$ is called an *ellipse*. If $a = b$ then this is a circle with *radius* $r = \frac{1}{a}$.

Sets of Points in the Plane
The *midpoint* between two points, (x_1, y_1) and (x_2, y_2), is given by taking the average of the *x* coordinates and the average of the *y* coordinates:

$$\left(\frac{x_1 + x_2}{2}, \frac{y_1 + y_2}{2} \right)$$

The *distance* between two points, (x_1, y_1) and (x_2, y_2), is given by the *Pythagorean formula*:

$$\sqrt{(x_2 - x_1)^2 + (y_2 - y_1)^2}$$

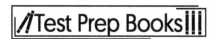
To find the perpendicular distance between a line $Ax + By = C$ and a point (x_1, y_1) not on the line, we need to use the formula:

$$\frac{|Ax_1 + By_1 + C|}{\sqrt{A^2 + B^2}}$$

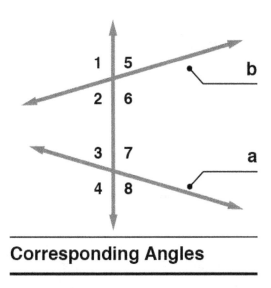

Corresponding Angles

Relationships between Angles

Supplementary angles add up to 180 degrees. *Vertical angles* are two nonadjacent angles formed by two intersecting lines. *Corresponding angles* are two angles in the same position whenever a straight line (known as a *transversal*) crosses two others. If the two lines are parallel, the corresponding angles are equal. *Alternate interior angles* are also a pair of angles formed when two lines are crossed by a transversal. They are opposite angles that exist inside of the two lines. In the corresponding angles diagram above, angles 2 and 7 are alternate interior angles, as well as angles 6 and 3. *Alternate exterior angles* are opposite angles formed by a transversal but, in contrast to interior angles, exterior angles exist outside the two original lines. Therefore, angles 1 and 8 are alternate exterior angles and so are angles 5 and 4. Finally, *consecutive interior angles* are pairs of angles formed by a transversal. These angles are located on the same side of the transversal and inside the two original lines. Therefore, angles 2 and 3 are a pair of consecutive interior angles, and so are angles 6 and 7. These definitions are instrumental in solving many problems that involve determining relationships between angles.

Medians, Midpoints, and Altitudes

A *median* of a triangle is the line drawn from a vertex to the midpoint on the opposite side. A triangle has three medians, and their point of intersection is known as the *centroid*. An *altitude* is a line drawn from a vertex perpendicular to the opposite side. A triangle has three altitudes, and their point of intersection is known as the *orthocenter*. An altitude can actually exist outside, inside, or on the triangle depending on the placement of the vertex. Many problems involve these definitions. For example, given one endpoint of a line segment and the midpoint, the other endpoint can be determined by using the midpoint formula. In addition, area problems heavily depend on these definitions. For example, it can be proven that the median of a triangle divides it into two regions of equal areas. The actual formula for the area of a triangle depends on its altitude.

Special Triangles

An *isosceles triangle* contains at least two equal sides. Therefore, it must also contain two equal angles and, subsequently, contain two medians of the same length. An isosceles triangle can also be labelled as an *equilateral triangle* (which contains three equal sides and three equal angles) when it meets these conditions. In an equilateral triangle, the measure of each angle is always 60 degrees. Also within an equilateral triangle, the medians are of the same length. A *scalene triangle* can never be an equilateral or an isosceles triangle because it contains no equal sides and no equal angles. Also, medians in a scalene triangle can't have the same length. However, a *right triangle*, which is a triangle containing a 90-degree angle, can be a scalene triangle. There are two types of special right triangles. The *30-60-90 right triangle* has angle measurements of 30 degrees, 60 degrees, and 90 degrees. Because of the nature of this triangle, and through the use of the Pythagorean theorem, the side lengths have a special relationship. If x is the length opposite the 30-degree angle, the length opposite the 60-degree angle is $\sqrt{3}x$, and the hypotenuse has length $2x$. The *45-45-90 right triangle* is also special as it contains two angle measurements of 45 degrees. It can be proven that, if x is the length of the two equal sides, the hypotenuse is $x\sqrt{2}$. The properties of all of these special triangles are extremely useful in determining both side lengths and angle measurements in problems where some of these quantities are given and some are not.

Special Quadrilaterals

A special quadrilateral is one in which both pairs of opposite sides are parallel. This type of quadrilateral is known as a *parallelogram*. A parallelogram has six important properties:

- Opposite sides are congruent.
- Opposite angles are congruent.
- Within a parallelogram, consecutive angles are supplementary, so their measurements total 180 degrees.
- If one angle is a right angle, all of them have to be right angles.
- The diagonals of the angles bisect each other.
- These diagonals form two congruent triangles.

A parallelogram with four congruent sides is a *rhombus*. A quadrilateral containing only one set of parallel sides is known as a *trapezoid*. The parallel sides are known as bases, and the other two sides are known as legs. If the legs are congruent, the trapezoid can be labelled an *isosceles trapezoid*. An important property of a trapezoid is that their diagonals are congruent. Also, the median of a trapezoid is parallel to the bases, and its length is equal to half of the sum of the base lengths.

Quadrilateral Relationships

Rectangles, squares, and rhombuses are *polygons* with four sides. By definition, all rectangles are parallelograms, but only some rectangles are squares. However, some parallelograms are rectangles. Also, it's true that all squares are rectangles, and some rhombuses are squares. There are no rectangles, squares, or rhombuses that are trapezoids though, because they have more than one set of parallel sides.

Diagonals and Angles

Diagonals are lines (excluding sides) that connect two vertices within a polygon. *Mutually bisecting diagonals* intersect at their midpoints. Parallelograms, rectangles, squares, and rhombuses have mutually bisecting diagonals. However, trapezoids don't have such lines. *Perpendicular diagonals* occur when they form four right triangles at their point of intersection. Squares and rhombuses have perpendicular diagonals, but trapezoids, rectangles, and parallelograms do not. Finally, *perpendicular bisecting* diagonals (also known as *perpendicular bisectors*) form four right triangles at their point of intersection, but this intersection is also the midpoint of the two lines. Both rhombuses and squares have perpendicular bisecting angles, but trapezoids, rectangles, and parallelograms do not. Knowing these definitions can help tremendously in problems that involve both angles and diagonals.

Polygons with More Than Four Sides

A *pentagon* is a five-sided figure. A six-sided shape is a *hexagon*. A seven-sided figure is classified as a *heptagon*, and an eight-sided figure is called an *octagon*. An important characteristic is whether a polygon is regular or irregular. If it's *regular*, the side lengths and angle measurements are all equal. An *irregular* polygon has unequal side lengths and angle measurements. Mathematical problems involving polygons with more than four sides usually involve side length and angle measurements. The sum of all internal angles in a polygon equals $180(n - 2)$ degrees, where n is the number of sides. Therefore, the total of all internal angles in a pentagon is 540 degrees because there are five sides so:

$$180(5 - 2) = 540 \text{ degrees}$$

Unfortunately, area formulas don't exist for polygons with more than four sides. However, their shapes can be split up into triangles, and the formula for area of a triangle can be applied and totaled to obtain the area for the entire figure.

Use Congruence and Similarity Criteria for Triangles to Solve Problems and Prove Geometric Relationships

Similar Figures and Proportions

Sometimes, two figures are similar, meaning they have the same basic shape and the same interior angles, but they have different dimensions. If the ratio of two corresponding sides is known, then that ratio, or scale factor, holds true for all of the dimensions of the new figure.

Here is an example of applying this principle. Suppose that Lara is 5 feet tall and is standing 30 feet from the base of a light pole, and her shadow is 6 feet long. How high is the light on the pole? To figure this, it helps to make a sketch of the situation:

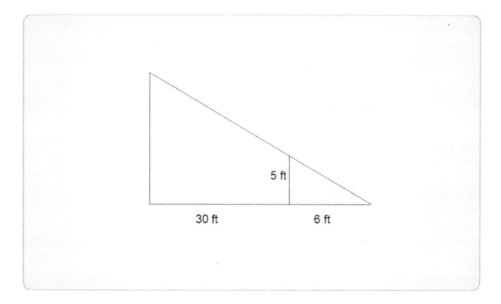

The light pole is the left side of the triangle. Lara is the 5-foot vertical line. Notice that there are two right triangles here, and that they have all the same angles as one another. Therefore, they form similar triangles. So, figure the ratio of proportionality between them.

The bases of these triangles are known. The small triangle, formed by Lara and her shadow, has a base of 6 feet. The large triangle, formed by the light pole along with the line from the base of the pole out to the end of Lara's shadow is $30 + 6 = 36$ feet long. So, the ratio of the big triangle to the little triangle will be $\frac{36}{6} = 6$. The height of the little triangle is 5 feet. Therefore, the height of the big triangle will be $6 \cdot 5 = 30$ feet, meaning that the light is 30 feet up the pole.

Notice that the perimeter of a figure changes by the ratio of proportionality between two similar figures, but the area changes by the *square* of the ratio. This is because if the length of one side is doubled, the area is quadrupled.

As an example, suppose two rectangles are similar, but the edges of the second rectangle are three times longer than the edges of the first rectangle. The area of the first rectangle is 10 square inches. How much more area does the second rectangle have than the first?

To answer this, note that the area of the second rectangle is $3^2 = 9$ times the area of the first rectangle, which is 10 square inches. Therefore, the area of the second rectangle is going to be $9 \cdot 10 = 90$ square inches. This means it has $90 - 10 = 80$ square inches more area than the first rectangle.

As a second example, suppose X and Y are similar right triangles. The hypotenuse of X is 4 inches. The area of Y is $\frac{1}{4}$ the area of X. What is the hypotenuse of Y?

First, realize the area has changed by a factor of $\frac{1}{4}$. The area changes by a factor that is the *square* of the ratio of changes in lengths, so the ratio of the lengths is the square root of the ratio of areas. That means that the ratio of lengths must be is $\sqrt{\frac{1}{4}} = \frac{1}{2}$, and the hypotenuse of *Y* must be $\frac{1}{2} \cdot 4 = 2$ inches.

Volumes between similar solids change like the cube of the change in the lengths of their edges. Likewise, if the ratio of the volumes between similar solids is known, the ratio between their lengths is known by finding the cube root of the ratio of their volumes.

For example, suppose there are two similar rectangular pyramids *X* and *Y*. The base of *X* is 1 inch by 2 inches, and the volume of *X* is 8 inches. The volume of *Y* is 64 inches. What are the dimensions of the base of *Y*?

To answer this, first find the ratio of the volume of *Y* to the volume of *X*. This will be given by $\frac{64}{8} = 8$. Now the ratio of lengths is the cube root of the ratio of volumes, or $\sqrt[3]{8} = 2$. So, the dimensions of the base of *Y* must be 2 inches by 4 inches.

Proving Geometric Theorems

Proving Theorems About Lines and Angles
To prove any geometric theorem, the proven theorems must be linked in a logical order that flows from an original point to the desired result. Proving theorems about lines and angles is the basis of proving theorems that involve other shapes. A *transversal* is a line that passes through two lines at two points. Common theorems that need to be proved are: vertical angles are congruent; a transversal passing through two parallel lines forms two alternate interior angles that are congruent and two corresponding angles that are congruent; and points on a perpendicular bisector of a line segment are equidistant from the endpoints of the line segment.

Triangle Theorems
To prove theorems about triangles, basic definitions involving triangles (e.g., equilateral, isosceles, etc.) need to be known. Proven theorems concerning lines and angles can be applied to prove theorems about triangles. Common theorems to be proved include: the sum of all angles in a triangle equals 180 degrees; the sum of the lengths of two sides of a triangle is greater than the length of the third side; the base angles of an isosceles triangle are congruent; the line segment connecting the midpoint of two sides of a triangle is parallel to the third side and its length is half the length of the third side; and the medians of a triangle all meet at a single point.

Parallelogram Theorems
A *parallelogram* is a quadrilateral with parallel opposing sides. Within parallelograms, opposite sides and angles are congruent and the diagonals bisect each other. Known theorems about parallel lines, transversals, complementary angles, and congruent triangles can be used to prove theorems about parallelograms. Theorems that need to be proved include: opposite sides of a parallelogram are congruent; opposite angles are congruent; the diagonals bisect each other; and rectangles are parallelograms with congruent diagonals.

Determining Congruence
Two figures are congruent if there is a rigid motion that can map one figure onto the other. Therefore, all pairs of sides and angles within the image and pre-image must be congruent. For example, in

triangles, each pair of the three sides and three angles must be congruent. Similarly, in two four-sided figures, each pair of the four sides and four angles must be congruent.

Similarity

Two figures are *similar* if there is a combination of translations, reflections, rotations, and dilations, which maps one figure onto the other. The difference between congruence and similarity is that dilation can be used in similarity. Therefore, side lengths between each shape can differ. However, angle measure must be preserved within this definition. If two polygons differ in size so that the lengths of corresponding line segments differ by the same factor, but corresponding angles have the same measurement, they are similar.

Triangle Congruence

There are five theorems to show that triangles are congruent when it's unknown whether each pair of angles and sides are congruent. Each theorem is a shortcut that involves different combinations of sides and angles that must be true for the two triangles to be congruent. For example, *side-side-side (SSS)* states that if all sides are equal, the triangles are congruent. *Side-angle-side (SAS)* states that if two pairs of sides are equal and the included angles are congruent, then the triangles are congruent. Similarly, *angle-side-angle (ASA)* states that if two pairs of angles are congruent and the included side lengths are equal, the triangles are similar. *Angle-angle-side (AAS)* states that two triangles are congruent if they have two pairs of congruent angles and a pair of corresponding equal side lengths that aren't included. Finally, *hypotenuse-leg (HL)* states that if two right triangles have equal hypotenuses and an equal pair of shorter sides, then the triangles are congruent. An important item to note is that angle-angle-angle *(AAA)* is not enough information to have congruence. It's important to understand why these rules work by using rigid motions to show congruence between the triangles with the given properties. For example, three reflections are needed to show why *SAS* follows from the definition of congruence.

Similarity for Two Triangles

If two angles of one triangle are congruent with two angles of a second triangle, the triangles are similar. This is because, within any triangle, the sum of the angle measurements is 180 degrees. Therefore, if two are congruent, the third angle must also be congruent because their measurements are equal. Three congruent pairs of angles mean that the triangles are similar.

Proving Congruence and Similarity

The criteria needed to prove triangles are congruent involves both angle and side congruence. Both pairs of related angles and sides need to be of the same measurement to use congruence in a proof. The criteria to prove similarity in triangles involves proportionality of side lengths. Angles must be congruent in similar triangles; however, corresponding side lengths only need to be a constant multiple of each other. Once similarity is established, it can be used in proofs as well. Relationships in geometric figures other than triangles can be proven using triangle congruence and similarity. If a similar or congruent triangle can be found within another type of geometric figure, their criteria can be used to prove a relationship about a given formula. For instance, a rectangle can be broken up into two congruent triangles.

Transformations of a Plane

Given a figure drawn on a plane, many changes can be made to that figure, including *rotation*, *translation*, and *reflection*. Rotations turn the figure about a point, translations slide the figure, and reflections flip the figure over a specified line. When performing these transformations, the original figure is called the *pre-image*, and the figure after transformation is called the *image*.

More specifically, *translation* means that all points in the figure are moved in the same direction by the same distance. In other words, the figure is slid in some fixed direction. Of course, while the entire figure is slid by the same distance, this does not change any of the measurements of the figures involved. The result will have the same distances and angles as the original figure.

In terms of Cartesian coordinates, a translation means a shift of each of the original points (x, y) by a fixed amount in the x and y directions, to become $(x + a, y + b)$.

Another procedure that can be performed is called *reflection*. To do this, a line in the plane is specified, called the *line of reflection*. Then, take each point and flip it over the line so that it is the same distance from the line but on the opposite side of it. This does not change any of the distances or angles involved, but it does reverse the order in which everything appears.

To reflect something over the *x*-axis, the points (x, y) are sent to $(x, -y)$. To reflect something over the *y*-axis, the points (x, y) are sent to the points $(-x, y)$. Flipping over other lines is not something easy to express in Cartesian coordinates. However, by drawing the figure and the line of reflection, the distance to the line and the original points can be used to find the reflected figure.

Example: Reflect this triangle with vertices (-1, 0), (2, 1), and (2, 0) over the *y*-axis. The pre-image is shown below.

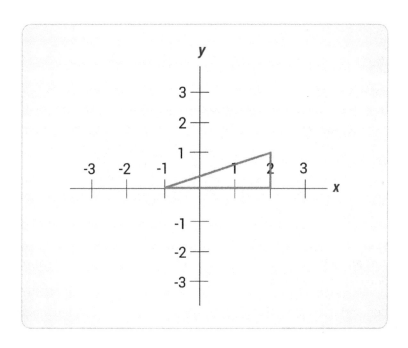

To do this, flip the *x* values of the points involved to the negatives of themselves, while keeping the *y* values the same. The image is shown here.

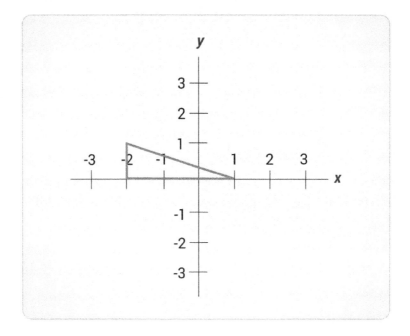

The new vertices will be (1, 0), (-2, 1), and (-2, 0).

Another procedure that does not change the distances and angles in a figure is *rotation*. In this procedure, pick a center point, then rotate every vertex along a circle around that point by the same angle. This procedure is also not easy to express in Cartesian coordinates, and this is not a requirement on this test. However, as with reflections, it's helpful to draw the figures and see what the result of the rotation would look like. This transformation can be performed using a compass and protractor.

Each one of these transformations can be performed on the coordinate plane without changes to the original dimensions or angles.

If two figures in the plane involve the same distances and angles, they are called *congruent figures*. In other words, two figures are congruent when they go from one form to another through reflection, rotation, and translation, or a combination of these.

Remember that rotation and translation will give back a new figure that is identical to the original figure, but reflection will give back a mirror image of it.

To recognize that a figure has undergone a rotation, check to see that the figure has not been changed into a mirror image, but that its orientation has changed (that is, whether the parts of the figure now form different angles with the *x* and *y* axes).

To recognize that a figure has undergone a translation, check to see that the figure has not been changed into a mirror image, and that the orientation remains the same.

To recognize that a figure has undergone a reflection, check to see that the new figure is a mirror image of the old figure.

Keep in mind that sometimes a combination of translations, reflections, and rotations may be performed on a figure.

Dilation

A *dilation* is a transformation that preserves angles, but not distances. This can be thought of as stretching or shrinking a figure. If a dilation makes figures larger, it is called an *enlargement*. If a dilation makes figures smaller, it is called a *reduction*. The easiest example is to dilate around the origin. In this case, multiply the *x* and *y* coordinates by a *scale factor*, k, sending points (x, y) to (kx, ky).

As an example, draw a dilation of the following triangle, whose vertices will be the points (-1, 0), (1, 0), and (1, 1).

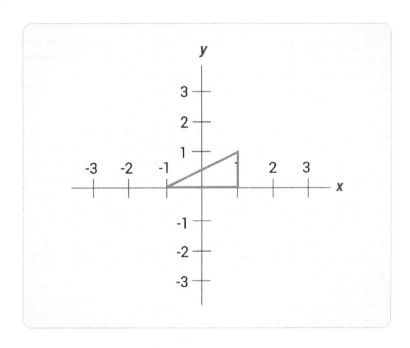

For this problem, dilate by a scale factor of 2, so the new vertices will be (-2, 0), (2, 0), and (2, 2).

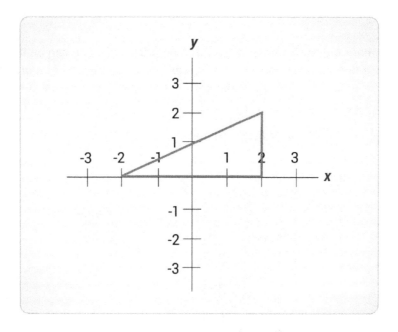

Note that after a dilation, the distances between the vertices of the figure will have changed, but the angles remain the same. The two figures that are obtained by dilation, along with possibly translation, rotation, and reflection, are all *similar* to one another. Another way to think of this is that similar figures have the same number of vertices and edges, and their angles are all the same. Similar figures have the same basic shape, but are different in size.

Symmetry

Using the types of transformations above, if an object can undergo these changes and not appear to have changed, then the figure is symmetrical. If an object can be split in half by a line and flipped over that line to lie directly on top of itself, it is said to have *line symmetry*. An example of both types of figures is seen below.

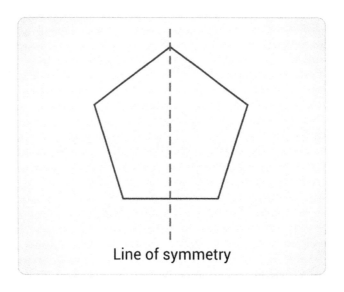

Line of symmetry

If an object can be rotated about its center to any degree smaller than 360, and it lies directly on top of itself, the object is said to have *rotational symmetry*. An example of this type of symmetry is shown below. The pentagon has an order of 5.

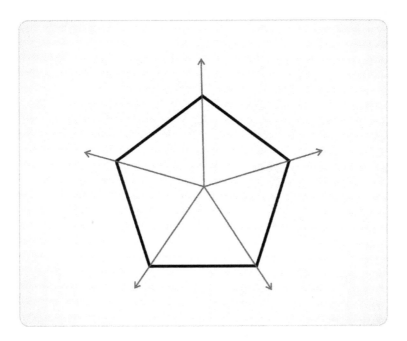

The rotational symmetry lines in the figure above can be used to find the angles formed at the center of the pentagon. Knowing that all of the angles together form a full circle, at 360 degrees, the figure can be split into 5 angles equally. By dividing the 360° by 5, each angle is 72°.

Given the length of one side of the figure, the perimeter of the pentagon can also be found using rotational symmetry. If one side length was 3 cm, that side length can be rotated onto each other side length four times. This would give a total of 5 side lengths equal to 3 cm. To find the perimeter, or distance around the figure, multiply 3 by 5. The perimeter of the figure would be 15 cm.

If a line cannot be drawn anywhere on the object to flip the figure onto itself or rotated less than or equal to 180 degrees to lay on top of itself, the object is asymmetrical. Examples of these types of figures are shown below.

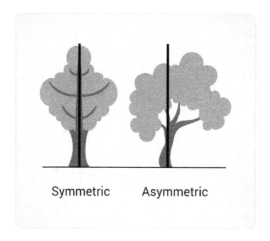

Symmetric   Asymmetric

No line of symmetry

Use Volume Formulas to Solve Problems

Geometry in three dimensions is similar to geometry in two dimensions. The main new feature is that three points now define a unique *plane* that passes through each of them. Three dimensional objects can be made by putting together two dimensional figures in different surfaces. Below, some of the possible three dimensional figures will be provided, along with formulas for their volumes and surface areas.

A rectangular prism is a box whose sides are all rectangles meeting at 90° angles. Such a box has three dimensions: length, width, and height. If the length is x, the width is y, and the height is z, then the volume is given by $V = xyz$.

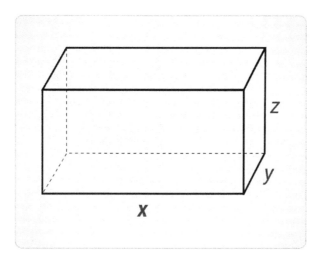

The surface area will be given by computing the surface area of each rectangle and adding them together. There are a total of six rectangles. Two of them have sides of length x and y, two have sides of length y and z, and two have sides of length x and z. Therefore, the total surface area will be given by:

$$SA = 2xy + 2yz + 2xz$$

A *rectangular pyramid* is a figure with a rectangular base and four triangular sides that meet at a single vertex. If the rectangle has sides of length x and y, then the volume will be given by:

$$V = \frac{1}{3}xyh$$

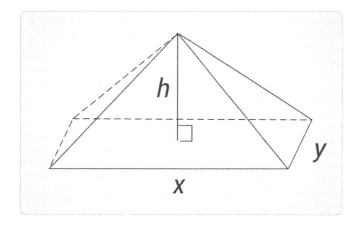

To find the surface area, the dimensions of each triangle need to be known. However, these dimensions can differ depending on the problem in question. Therefore, there is no general formula for calculating total surface area.

A *sphere* is a set of points all of which are equidistant from some central point. It is like a circle, but in three dimensions. The volume of a sphere of radius *r* is given by:

$$V = \frac{4}{3}\pi r^3$$

The surface area is given by:

$$A = 4\pi r^2$$

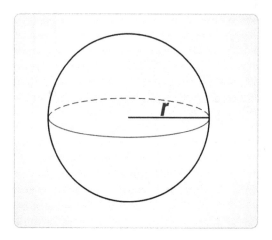

Apply Concepts of Density Based on Area and Volume in Modeling Situations

Real-World Geometry

Many real-world objects can be compared to geometric shapes. Describing certain objects using the measurements and properties of two- and three-dimensional shapes is an important part of geometry. For example, basic ideas such as angles and line segments can be seen in real-world objects. The corner of any room is an angle, and the intersection of a wall with the floor is like a line segment. Building upon this idea, entire objects can be related to both two- and three-dimensional shapes. An entire room can be thought of as square, rectangle, or a sum of a few three-dimensional shapes. Knowing what properties and measures are needed to make decisions in real life is why geometry is such a useful branch of mathematics. One obvious relationship between a real-life situation and geometry exists in construction. For example, to build an addition onto a house, several geometric measurements will be used.

Density

The *density* of a substance is the ratio of mass to area or volume. It's a relationship between the mass and how much space the object actually takes up. Knowing which units to use in each situation is crucial. Population density is an example of a real-life situation that's modeled by using density concepts. It involves calculating the ratio of the number of people to the number of square miles. The amount of material needed per a specific unit of area or volume is another application. For example, estimating the number of BTUs per cubic foot of a home is a measurement that relates to heating or cooling the house based on the desired temperature and the house's size.

Solving Design Problem

Design problems are an important application of geometry (e.g., building structures that satisfy physical constraints and/or minimize costs). These problems involve optimizing a situation based on what's given and required. For example, determining what size barn to build, given certain dimensions and a specific budget, uses both geometric properties and other mathematical concepts. Equations are formed using geometric definitions and the given constraints. In the end, such problems involve solving a system of equations and rely heavily on a strong background in algebra. *Typographic grid systems* also help with such design problems. A grid made up of intersecting straight or curved lines can be used as a visual representation of the structure being designed. This concept is seen in the blueprints used throughout the graphic design process.

The Pythagorean Theorem

The Pythagorean theorem is an important result in geometry. It states that for right triangles, the sum of the squares of the two shorter sides will be equal to the square of the longest side (also called the *hypotenuse*). The longest side will always be the side opposite to the 90° angle. If this side is called c, and the other two sides are a and b, then the Pythagorean theorem states that:

$$c^2 = a^2 + b^2$$

Since lengths are always positive, this also can be written as:

$$c = \sqrt{a^2 + b^2}$$

A diagram to show the parts of a triangle using the Pythagorean theorem is below.

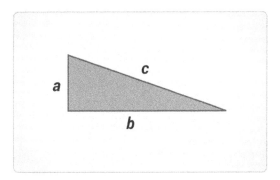

As an example of the theorem, suppose that Shirley has a rectangular field that is 5 feet wide and 12 feet long, and she wants to split it in half using a fence that goes from one corner to the opposite corner. How long will this fence need to be? To figure this out, note that this makes the field into two right triangles, whose hypotenuse will be the fence dividing it in half. Therefore, the fence length will be given by:

$$\sqrt{5^2 + 12^2} = \sqrt{169} = 13 \text{ feet long}$$

Numbers and Quantity

Rewrite Expressions Involving Radicals and Rational Exponents Using the Properties of Exponents

Properties of Exponents

Exponents are used in mathematics to express a number or variable multiplied by itself a certain number of times. For example, x^3 means x is multiplied by itself three times. In this expression, x is called the *base*, and 3 is the *exponent*. Exponents can be used in more complex problems when they contain fractions and negative numbers.

Fractional exponents can be explained by looking first at the inverse of exponents, which are *roots*. Given the expression x^2, the square root can be taken, $\sqrt{x^2}$, cancelling out the 2 and leaving x by itself, if x is positive. Cancellation occurs because $\sqrt{x}$ can be written with exponents, instead of roots, as $x^{\frac{1}{2}}$. The numerator of 1 is the exponent, and the denominator of 2 is called the root (which is why it's referred to as *square root*). Taking the square root of x^2 is the same as raising it to the $\frac{1}{2}$ power. Written out in mathematical form, it takes the following progression:

$$\sqrt{x^2} = (x^2)^{\frac{1}{2}} = x$$

From properties of exponents, $2 \cdot \frac{1}{2} = 1$ is the actual exponent of x. Another example can be seen with $x^{\frac{4}{7}}$. The variable *x,* raised to four-sevenths, is equal to the seventh root of x to the fourth power: $\sqrt[7]{x^4}$. In general,

$$x^{\frac{1}{n}} = \sqrt[n]{x}$$

and

$$x^{\frac{m}{n}} = \sqrt[n]{x^m}$$

Negative exponents also involve fractions. Whereas y^3 can also be rewritten as $\frac{y^3}{1}$, y^{-3} can be rewritten as $\frac{1}{y^3}$. A negative exponent means the exponential expression must be moved to the opposite spot in a fraction to make the exponent positive. If the negative appears in the numerator, it moves to the denominator. If the negative appears in the denominator, it is moved to the numerator. In general, $a^{-n} = \frac{1}{a^n}$, and a^{-n} and a^n are reciprocals.

Take, for example, the following expression:

$$\frac{a^{-4}b^2}{c^{-5}}$$

Since *a* is raised to the negative fourth power, it can be moved to the denominator. Since *c* is raised to the negative fifth power, it can be moved to the numerator. The *b* variable is raised to the positive second power, so it does not move.

The simplified expression is as follows:

$$\frac{b^2 c^5}{a^4}$$

In mathematical expressions containing exponents and other operations, the order of operations must be followed. *PEMDAS* states that exponents are calculated after any parenthesis and grouping symbols, but before any multiplication, division, addition, and subtraction.

The Evaluation of Positive Rational Roots and Exponents

There are a few rules for working with exponents. For any numbers a, b, m, n, the following hold true:

$$a^1 = a$$

$$1^a = 1$$

$$a^0 = 1$$

$$a^m \times a^n = a^{m+n}$$

$$a^m \div a^n = a^{m-n}$$

$$(a^m)^n = a^{m \times n}$$

$$(a \times b)^m = a^m \times b^m$$

$$(a \div b)^m = a^m \div b^m$$

Any number, including a fraction, can be an exponent. The same rules apply.

Manipulating Roots and Exponents

A *root* is a different way to write an exponent when the exponent is the reciprocal of a whole number. We use the *radical* symbol to write this in the following way: $\sqrt[n]{a} = a^{\frac{1}{n}}$. This quantity is called the *n-th root* of a. The n is called the *index* of the radical.

Note that if the *n*-th root of a is multiplied by itself n times, the result will just be a. If no number n is written by the radical, it is assumed that n is 2: $\sqrt{5} = 5^{\frac{1}{2}}$. The special case of the 2nd root is called the *square root*, and the third root is called the *cube root*.

A *perfect square* is a whole number that is the square of another whole number. For example, sixteen and 64 are perfect squares because 16 is the square of 4, and 64 is the square of 8.

Properties of Rational and Irrational Numbers

All real numbers can be separated into two groups: rational and irrational numbers. *Rational numbers* are any numbers that can be written as a fraction, such as $\frac{1}{3}, \frac{7}{4}$, and -25. Alternatively, *irrational numbers* are those that cannot be written as a fraction, such as numbers with never-ending, non-repeating decimal values. Many irrational numbers result from taking roots, such as $\sqrt{2}$ or $\sqrt{3}$. An irrational number may be written as:

$$34.5684952\ldots$$

The ellipsis (…) represents the line of numbers after the decimal that does not repeat and is never-ending.

When rational and irrational numbers interact, there are different types of number outcomes. For example, when adding or multiplying two rational numbers, the result is a rational number. No matter what two fractions are added or multiplied together, the result can always be written as a fraction. The following expression shows two rational numbers multiplied together:

$$\frac{3}{8} * \frac{4}{7} = \frac{12}{56}$$

The product of these two fractions is another fraction that can be simplified to $\frac{3}{14}$.

As another interaction, rational numbers added to irrational numbers will always result in irrational numbers. No part of any fraction can be added to a never-ending, non-repeating decimal to make a rational number. The same result is true when multiplying a rational and irrational number. Taking a fractional part of a never-ending, non-repeating decimal will always result in another never-ending, non-repeating decimal. An example of the product of rational and irrational numbers is shown in the following expression: $2 * \sqrt{7}$.

The last type of interaction concerns two irrational numbers, where the sum or product may be rational or irrational depending on the numbers being used. The following expression shows a rational sum from two irrational numbers:

$$\sqrt{3} + \left(6 - \sqrt{3}\right) = 6$$

The product of two irrational numbers can be rational or irrational. A rational result can be seen in the following expression:

$$\sqrt{2} * \sqrt{8} = \sqrt{2 * 8} = \sqrt{16} = 4$$

An irrational result can be seen in the following:

$$\sqrt{3} * \sqrt{2} = \sqrt{6}$$

Strategies and Algorithms to Perform Operations on Rational Numbers

A rational number is any number that can be written in the form of a ratio or fraction. Integers can be written as fractions with a denominator of 1 ($5 = \frac{5}{1}$; $-342 = \frac{-342}{1}$; etc.).

Decimals that terminate and/or repeat can also be written as fractions ($47 = \frac{47}{100}$; $.\overline{33} = \frac{1}{3}$).

When adding or subtracting fractions, the numbers must have the same denominators. In these cases, numerators are added or subtracted and denominators are kept the same. For example, $\frac{2}{7} + \frac{3}{7} = \frac{5}{7}$ and $\frac{4}{5} - \frac{3}{5} = \frac{1}{5}$. If the fractions to be added or subtracted do not have the same denominator, a common denominator must be found. This is accomplished by changing one or both fractions to a different but equivalent fraction. Consider the example $\frac{1}{6} + \frac{4}{9}$. First, a common denominator must be found. One method is to find the least common multiple (LCM) of the denominators 6 and 9. This is the lowest number that both 6 and 9 will divide into evenly. In this case the LCM is 18. Both fractions should be

changed to equivalent fractions with a denominator of 18. To obtain the numerator of the new fraction, the old numerator is multiplied by the same number by which the old denominator is multiplied. For the fraction $\frac{1}{6}$, 6 multiplied by 3 will produce a denominator of 18. Therefore, the numerator is multiplied by 3 to produce the new numerator $\left(\frac{1\times3}{6\times3} = \frac{3}{18}\right)$. For the fraction $\frac{4}{9}$, multiplying both the numerator and denominator by 2 produces $\frac{8}{18}$. Since the two new fractions have common denominators, they can be added:

$$\left(\frac{3}{18} + \frac{8}{18} = \frac{11}{18}\right)$$

When multiplying or dividing rational numbers, these numbers may be converted to fractions and multiplied or divided accordingly. When multiplying fractions, all numerators are multiplied by each other and all denominators are multiplied by each other. For example:

$$\frac{1}{3}\times\frac{6}{5} = \frac{1\times6}{3\times5} = \frac{6}{15}\text{ and }\frac{-1}{2}\times\frac{3}{1}\times\frac{11}{100} = \frac{-1\times3\times11}{2\times1\times100} = \frac{-33}{200}$$

When dividing fractions, the problem is converted by multiplying by the reciprocal of the divisor. This is done by changing division to multiplication and "flipping" the second fraction, or divisor. For example,

$$\frac{1}{2}\div\frac{3}{5}\rightarrow\frac{1}{2}\times\frac{5}{3}\text{ and }\frac{5}{1}\div\frac{1}{3}\rightarrow\frac{5}{1}\times\frac{3}{1}$$

To complete the problem, the rules for multiplying fractions should be followed.

Note that when adding, subtracting, multiplying, and dividing mixed numbers (ex. $4\frac{1}{2}$), it is easiest to convert these to improper fractions (larger numerator than denominator). To do so, the denominator is kept the same. To obtain the numerator, the whole number is multiplied by the denominator and added to the numerator. For example, $4\frac{1}{2} = \frac{9}{2}$ and $7\frac{2}{3} = \frac{23}{3}$. Also, note that answers involving fractions should be converted to the simplest form.

Use Units as a Way to Understand Problems and Guide the Solution

Ratios and Proportions

Ratios are used to show the relationship between two quantities. The ratio of oranges to apples in the grocery store may be 3 to 2. That means that for every 3 oranges, there are 2 apples. This comparison can be expanded to represent the actual number of oranges and apples. Another example may be the number of boys to girls in a math class. If the ratio of boys to girls is given as 2 to 5, that means there are 2 boys to every 5 girls in the class. Ratios can also be compared if the units in each ratio are the same. The ratio of boys to girls in the math class can be compared to the ratio of boys to girls in a science class by stating which ratio is higher and which is lower.

Rates are used to compare two quantities with different units. *Unit rates* are the simplest form of rate. With unit rates, the denominator in the comparison of two units is one. For example, if someone can type at a rate of 1000 words in 5 minutes, then his or her unit rate for typing is $\frac{1000}{5} = 200$ words in one minute or 200 words per minute. Any rate can be converted into a unit rate by dividing to make the denominator one. 1000 words in 5 minutes has been converted into the unit rate of 200 words per minute.

Ratios and rates can be used together to convert rates into different units. For example, if someone is driving 50 kilometers per hour, that rate can be converted into miles per hour by using a ratio known as the *conversion factor*. Since the given value contains kilometers and the final answer needs to be in miles, the ratio relating miles to kilometers needs to be used. There are 0.62 miles in 1 kilometer. This, written as a ratio and in fraction form, is:

$$\frac{0.62 \ miles}{1 \ km}$$

To convert 50km/hour into miles per hour, the following conversion needs to be set up:

$$\frac{50 \ km}{hour} * \frac{0.62 \ miles}{1 \ km} = 31 \ miles \ per \ hour$$

The ratio between two similar geometric figures is called the *scale factor*. For example, a problem may depict two similar triangles, A and B. The scale factor from the smaller triangle A to the larger triangle B is given as 2 because the length of the corresponding side of the larger triangle, 16, is twice the corresponding side on the smaller triangle, 8. This scale factor can also be used to find the value of a missing side, x, in triangle A. Since the scale factor from the smaller triangle (A) to larger one (B) is 2, the larger corresponding side in triangle B (given as 25), can be divided by 2 to find the missing side in A ($x = 12.5$). The scale factor can also be represented in the equation $2A = B$ because two times the lengths of A gives the corresponding lengths of B. This is the idea behind similar triangles.

Much like a scale factor can be written using an equation like $2A = B$, a *relationship* is represented by the equation $Y = kX$. X and Y are proportional because as values of X increase, the values of Y also increase. A relationship that is inversely proportional can be represented by the equation $Y = \frac{k}{x}$, where the value of Y decreases as the value of x increases and vice versa.

Proportional reasoning can be used to solve problems involving ratios, percentages, and averages. Ratios can be used in setting up proportions and solving them to find unknowns. For example, if a student completes an average of 10 pages of math homework in 3 nights, how long would it take the student to complete 22 pages? Both ratios can be written as fractions. The second ratio would contain the unknown.

The following proportion represents this problem, where x is the unknown number of nights:

$$\frac{10 \ pages}{3 \ nights} = \frac{22 \ pages}{x \ nights}$$

Solving this proportion entails cross-multiplying and results in the following equation: $10x = 22 * 3$. Simplifying and solving for x results in the exact solution: $x = 6.6 \ nights$. The result would be rounded up to 7 because the homework would actually be completed on the 7th night.

The following problem uses ratios involving percentages:

If 20% of the class is girls and 30 students are in the class, how many girls are in the class?

To set up this problem, it is helpful to use the common proportion:

$$\frac{\%}{100} = \frac{is}{of}$$

Within the proportion, % is the percentage of girls, 100 is the total percentage of the class, *is* is the number of girls, and *of* is the total number of students in the class. Most percentage problems can be written using this language. To solve this problem, the proportion should be set up as $\frac{20}{100} = \frac{x}{30}$, and then solved for x. Cross-multiplying results in the equation $20 * 30 = 100x$, which results in the solution $x = 6$. There are 6 girls in the class.

Problems involving volume, length, and other units can also be solved using ratios. A problem may ask for the volume of a cone to be found that has a radius, $r = 7m$ and a height, $h = 16m$. Referring to the formulas provided on the test, the volume of a cone is given as:

$$V = \pi r^2 \frac{h}{3}$$

r is the radius, and h is the height. Plugging $r = 7$ and $h = 16$ into the formula, the following is obtained:

$$V = \pi(7^2)\frac{16}{3}$$

Therefore, volume of the cone is found to be approximately 821m³. Sometimes, answers in different units are sought. If this problem wanted the answer in liters, 821m³ would need to be converted.

Using the equivalence statement 1m³ = 1000L, the following ratio would be used to solve for liters:

$$821\text{m}^3 * \frac{1000L}{1m^3}$$

Cubic meters in the numerator and denominator cancel each other out, and the answer is converted to 821,000 liters, or $8.21 * 10^5$ L.

Other conversions can also be made between different given and final units. If the temperature in a pool is 30°C, what is the temperature of the pool in degrees Fahrenheit? To convert these units, an equation is used relating Celsius to Fahrenheit. The following equation is used:

$$T_{\circ F} = 1.8T_{\circ C} + 32$$

Plugging in the given temperature and solving the equation for T yields the result:

$$T_{\circ F} = 1.8(30) + 32 = 86°F$$

Both units in the metric system and U.S. customary system are widely used.

Solving Real-World Problems Involving Proportions

Much like a scale factor can be written using an equation like $2A = B$, a *relationship* is represented by the equation $Y = kX$. X and Y are proportional because as values of X increase, the values of Y also increase. A relationship that is inversely proportional can be represented by the equation $Y = \frac{k}{X}$, where the value of Y decreases as the value of x increases and vice versa.

Proportional reasoning can be used to solve problems involving ratios, percentages, and averages. Ratios can be used in setting up proportions and solving them to find unknowns. For example, if a student completes an average of 10 pages of math homework in 3 nights, how long would it take the student to complete 22 pages? Both ratios can be written as fractions. The second ratio would contain the unknown.

The following proportion represents this problem, where x is the unknown number of nights:

$$\frac{10 \; pages}{3 \; nights} = \frac{22 \; pages}{x \; nights}$$

Solving this proportion entails cross-multiplying and results in the following equation: $10x = 22 * 3$. Simplifying and solving for x results in the exact solution: $x = 6.6 \; nights$. The result would be rounded up to 7 because the homework would actually be completed on the 7th night.

The following problem uses ratios involving percentages:

If 20% of the class is girls and 30 students are in the class, how many girls are in the class?

To set up this problem, it is helpful to use the common proportion:

$$\frac{\%}{100} = \frac{is}{of}$$

Within the proportion, % is the percentage of girls, 100 is the total percentage of the class, *is* is the number of girls, and *of* is the total number of students in the class. Most percentage problems can be written using this language. To solve this problem, the proportion should be set up as $\frac{20}{100} = \frac{x}{30}$, and then solved for x. Cross-multiplying results in the equation $20 * 30 = 100x$, which results in the solution $x = 6$. There are 6 girls in the class.

Problems involving volume, length, and other units can also be solved using ratios. For example, a problem may ask for the volume of a cone to be found that has a radius, $r = 7m$ and a height, $h = 16m$. Referring to the formulas provided on the test, the volume of a cone is given as: $V = \pi r^2 \frac{h}{3}$, where r is the radius, and h is the height. Plugging $r = 7$ and $h = 16$ into the formula, the following is obtained:

$$V = \pi(7^2)\frac{16}{3}$$

Therefore, volume of the cone is found to be approximately 821m³. Sometimes, answers in different units are sought. If this problem wanted the answer in liters, 821m³ would need to be converted. Using the equivalence statement 1m³ = 1000L, the following ratio would be used to solve for liters:

$$821\text{m}^3 * \frac{1000L}{1m^3}$$

Cubic meters in the numerator and denominator cancel each other out, and the answer is converted to 821,000 liters, or $8.21 * 10^5$ L.

Other conversions can also be made between different given and final units. If the temperature in a pool is 30°C, what is the temperature of the pool in degrees Fahrenheit? To convert these units, an equation is used relating Celsius to Fahrenheit. The following equation is used:

$$T_{°F} = 1.8T_{°C} + 32$$

Plugging in the given temperature and solving the equation for T yields the result:

$$T_{\circ F} = 1.8(30) + 32 = 86^\circ F$$

Both units in the metric system and U.S. customary system are widely used.

Here are some more examples of how to solve for proportions:

1) $\dfrac{75\%}{90\%} = \dfrac{25\%}{x}$

To solve for x, the fractions must be cross multiplied:

$$(75\%x = 90\% \times 25\%)$$

To make things easier, let's convert the percentages to decimals:

$$(0.9 \times 0.25 = 0.225 = 0.75x)$$

To get rid of x's co-efficient, each side must be divided by that same coefficient to get the answer $x = 0.3$. The question could ask for the answer as a percentage or fraction in lowest terms, which are 30% and $\dfrac{3}{10}$, respectively.

2) $\dfrac{x}{12} = \dfrac{30}{96}$

Cross-multiply: $96x = 30 \times 12$

Multiply: $96x = 360$

Divide: $x = 360 \div 96$

Answer: $x = 3.75$

3) $\dfrac{0.5}{3} = \dfrac{x}{6}$

Cross-multiply: $3x = 0.5 \times 6$

Multiply: $3x = 3$

Divide: $x = 3 \div 3$

Answer: $x = 1$

You may have noticed there's a faster way to arrive at the answer. If there is an obvious operation being performed on the proportion, the same operation can be used on the other side of the proportion to solve for x. For example, in the first practice problem, 75% became 25% when divided by 3, and upon doing the same to 90%, the correct answer of 30% would have been found with much less legwork. However, these questions aren't always so intuitive, so it's a good idea to work through the steps, even if the answer seems apparent from the outset.

Solving Single- and Multistep Problems Involving Measurement Quantities, Units, and Unit Conversion

Unit Rates

A rate is a ratio in which two terms are in different units. When rates are expressed as a quantity of one, they are considered unit rates. To determine a unit rate, the first quantity is divided by the second. Knowing a unit rate makes calculations easier than simply having a rate. Suppose someone bought a 3lb bag of onions for $1.77. To calculate the price of 5lbs of onions, a proportion could be set up as follows:

$$\frac{3}{1.77} = \frac{5}{x}$$

However, knowing the unit rate, multiplying the value of pounds of onions by the unit price is another way to find the solution: (The unit price would be calculated $1.77/3lb = $0.59/lb.)

$$5lbs \times \frac{\$.59}{lb} = \$2.95 \text{ (The "lbs" units cancel out.)}$$

Unit Conversion

Unit conversions apply to many real-world scenarios, including cooking, measurement, construction, and currency. Problems on this material can be solved similarly to those involving unit rates. Given the conversion rate, it can be written as a fraction (ratio) and multiplied by a quantity in one unit to convert it to the corresponding unit. For example, someone might want to know how many minutes are in 3½ hours. The conversion rate of 60 minutes to 1 hour can be written as $\frac{60\ min}{1\ h}$. Multiplying the quantity by the conversion rate results in:

$$3\frac{1}{2}h \times \frac{60\ min}{1\ h} = 210\ min$$

The "h" unit is canceled. To convert a quantity in minutes to hours, the fraction for the conversion rate would be flipped (to cancel the "min" unit). To convert 195 minutes to hours, the equation:

$$195\ min \times \frac{1h}{60min} \text{ would be used}$$

The result is $\frac{195h}{60}$, which reduces to $3\frac{1}{4}$ hours.

Converting units may require more than one multiplication. The key is to set up the conversion rates so that units cancel out each other and the desired unit is left. Suppose someone wants to convert 3.25 yards to inches, given that 1yd = 3ft and 12in = 1ft. To calculate, the equation:

$$3.25yd \times \frac{3ft}{1yd} \times \frac{12in}{1ft} \text{ would be used}$$

The "yd" and "ft" units will cancel, resulting in 117 inches.

Solving Real Word One- or Multi-Step Problems with Rational Numbers

Word problems can appear daunting, but don't let the verbiage psych you out. No matter the scenario or specifics, the key to answering them is to translate the words into a math problem. Always keep in mind what the question is asking and what operations could lead to that answer. The following word problems highlight the most commonly tested question types.

Working with Money
Walter's Coffee Shop sells a variety of drinks and breakfast treats.

Price List	
Hot Coffee	$2.00
Slow-Drip Iced Coffee	$3.00
Latte	$4.00
Muffin	$2.00
Crepe	$4.00
Egg Sandwich	$5.00

Costs	
Hot Coffee	$0.25
Slow-Drip Iced Coffee	$0.75
Latte	$1.00
Muffin	$1.00
Crepe	$2.00
Egg Sandwich	$3.00

Walter's utilities, rent, and labor costs him $500 per day. Today, Walter sold 200 hot coffees, 100 slow-drip iced coffees, 50 lattes, 75 muffins, 45 crepes, and 60 egg sandwiches. What was Walter's total profit today?

To accurately answer this type of question, determine the total cost of making his drinks and treats, then determine how much revenue he earned from selling those products. After arriving at these two totals, the profit is measured by deducting the total cost from the total revenue.

Walter's costs for today:

Item	Quantity	Cost Per Unit	Total Cost
Hot Coffee	200	$0.25	$50
Slow-Drip Iced Coffee	100	$0.75	$75
Latte	50	$1.00	$50
Muffin	75	$1.00	$75
Crepe	45	$2.00	$90
Egg Sandwich	60	$3.00	$180
Utilities, rent, and labor			$500
Total Costs			$1,020

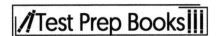

Walter's revenue for today:

Item	Quantity	Revenue Per Unit	Total Revenue
Hot Coffee	200	$2.00	$400
Slow-Drip Iced Coffee	100	$3.00	$300
Latte	50	$4.00	$200
Muffin	75	$2.00	$150
Crepe	45	$4.00	$180
Egg Sandwich	60	$5.00	$300
Total Revenue			$1,530

Walter's Profit = *Revenue – Costs* = $1,530 – $1,020 = $510

This strategy is applicable to other question types. For example, calculating salary after deductions, balancing a checkbook, and calculating a dinner bill are common word problems similar to business planning. Just remember to use the correct operations. When a balance is increased, use addition. When a balance is decreased, use subtraction. Common sense and organization are your greatest assets when answering word problems.

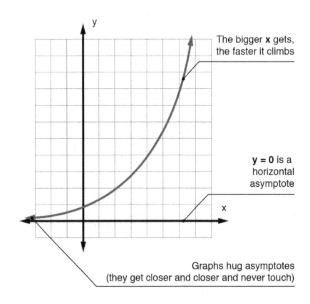

Percentages

Think of percentages as fractions with a denominator of 100. In fact, percentage means "per hundred." Problems often require converting numbers from percentages, fractions, and decimals.

Percent Problems

The basic percent equation is the following:

$$\frac{is}{of} = \frac{\%}{100}$$

The placement of numbers in the equation depends on what the question asks.

Example 1
Find 40% of 80.

Basically, the problem is asking, "What is 40% of 80?" The 40% is the percent, and 80 is the number to find the percent "of." The equation is:

$$\frac{x}{80} = \frac{40}{100}$$

Solving the equation by cross-multiplication, the problem becomes 100x = 80(40). Solving for x gives the answer: x = 32.

Example 2
What percent of 100 is 20?

The 20 fills in the "is" portion, while 100 fills in the "of." The question asks for the percent, so that will be x, the unknown. The following equation is set up:

$$\frac{20}{100} = \frac{x}{100}$$

Cross-multiplying yields the equation 100x = 20(100). Solving for x gives the answer of 20%.

Example 3
30% of what number is 30?

The following equation uses the clues and numbers in the problem:

$$\frac{30}{x} = \frac{30}{100}$$

Cross-multiplying results in the equation 30(100) = 30x. Solving for x gives the answer x = 100.

Decimals and Percentages

Since a percentage is based on "per hundred," decimals and percentages can be converted by multiplying or dividing by 100. Practically speaking, this always amounts to moving the decimal point two places to the right or left, depending on the conversion. To convert a percentage to a decimal, move the decimal point two places to the left and remove the % sign. To convert a decimal to a percentage, move the decimal point two places to the right and add a "%" sign. Here are some examples:

65% = 0.65
0.33 = 33%
0.215 = 21.5%
99.99% = 0.9999
500% = 5.00
7.55 = 755%

Fractions and Percentages

Remember that a percentage is a number per one hundred. So, a percentage can be converted to a fraction by making the number in the percentage the numerator and putting 100 as the denominator:

$$43\% = \frac{43}{100}$$

$$97\% = \frac{97}{100}$$

Note that the percent symbol (%) kind of looks like a 0, a 1, and another 0. So, think of a percentage like 54% as 54 over 100.

To convert a fraction to a percent, follow the same logic. If the fraction happens to have 100 in the denominator, you're in luck. Just take the numerator and add a percent symbol:

$$\frac{28}{100} = 28\%$$

Otherwise, divide the numerator by the denominator to get a decimal:

$$\frac{9}{12} = 0.75$$

Then convert the decimal to a percentage:

$$0.75 = 75\%$$

Another option is to make the denominator equal to 100. Be sure to multiply the numerator by the same number as the denominator. For example:

$$\frac{3}{20} \times \frac{5}{5} = \frac{15}{100}$$

$$\frac{15}{100} = 15\%$$

Changing Fractions to Decimals

To change a fraction into a decimal, divide the denominator into the numerator until there are no remainders. There may be repeating decimals, so rounding is often acceptable. A straight line above the repeating portion denotes that the decimal repeats.

Example

Express 4/5 as a decimal.

Set up the division problem.

$$5\overline{)4}$$

5 does not go into 4, so place the decimal and add a zero.

$$5\overline{)4.0}$$

5 goes into 40 eight times. There is no remainder.

$$\begin{array}{r} 0.8 \\ 5\overline{)4.0} \\ -\ 4.0 \\ \hline 0 \end{array}$$

The solution is 0.8.

Example

Express 33 1/3 as a decimal.

Since the whole portion of the number is known, set it aside to calculate the decimal from the fraction portion.

Set up the division problem.

$$3\overline{)1}$$

3 does not go into 1, so place the decimal and add zeros. 3 goes into 10 three times.

$$\begin{array}{r} 0.333 \\ 3\overline{)1.000} \end{array}$$

This will repeat with a remainder of 1, so place a line over the 3 denotes the repetition.

$$
\begin{array}{r}
0.333 \\
3\overline{)1.000} \\
-9 \\
\hline
10 \\
-9 \\
\hline
10
\end{array}
$$

The solution is $0.\overline{3}$

Changing Decimals to Fractions

To change decimals to fractions, place the decimal portion of the number, the numerator, over the respective place value, the denominator, then reduce, if possible.

Example

Express 0.25 as a fraction.

This is read as twenty-five hundredths, so put 25 over 100. Then reduce to find the solution.

$$
\frac{25}{100} = \frac{1}{4}
$$

Example

Express 0.455 as a fraction

This is read as four hundred fifty-five thousandths, so put 455 over 1000. Then reduce to find the solution.

$$
\frac{455}{1000} = \frac{91}{200}
$$

There are two types of problems that commonly involve percentages. The first is to calculate some percentage of a given quantity, where you convert the percentage to a decimal, and multiply the quantity by that decimal. Secondly, you are given a quantity and told it is a fixed percent of an unknown quantity. In this case, convert to a decimal, then divide the given quantity by that decimal.

Example

What is 30% of 760?

Convert the percent into a useable number. "Of" means to multiply.

$$
30\% = 0.30
$$

Set up the problem based on the givens and solve.

$$
0.30 \times 760 = 228
$$

Example
8.4 is 20% of what number?

Convert the percent into a useable number.

$$20\% = 0.20$$

The given number is a percent of the answer needed, so divide the given number by this decimal rather than multiplying it.

$$\frac{8.4}{0.20} = 42$$

Solving Real-World Problems Involving Percentages
Questions dealing with percentages can be difficult when they are phrased as word problems. These word problems almost always come in three varieties. The first type will ask to find what percentage of some number will equal another number. The second asks to determine what number is some percentage of another given number. The third will ask what number another number is a given percentage of.

One of the most important parts of correctly answering percentage word problems is to identify the numerator and the denominator. This fraction can then be converted into a percentage, as described above.

The following word problem shows how to make this conversion:

A department store carries several different types of footwear. The store is currently selling 8 athletic shoes, 7 dress shoes, and 5 sandals. What percentage of the store's footwear are sandals?

First, calculate what serves as the 'whole', as this will be the denominator. How many total pieces of footwear does the store sell? The store sells 20 different types (8 athletic + 7 dress + 5 sandals).

Second, what footwear type is the question specifically asking about? Sandals. Thus, 5 is the numerator.

Third, the resultant fraction must be expressed as a percentage. The first two steps indicate that $\frac{5}{20}$ of the footwear pieces are sandals. This fraction must now be converted into a percentage:

$$\frac{5}{20} \times \frac{5}{5} = \frac{25}{100} = 25\%$$

Solving for X in Proportions
Proportions are commonly used to solve word problems to find unknown values such as x that are some percent or fraction of a known number. Proportions are solved by cross-multiplying and then dividing to arrive at x. The following examples show how this is done:

1) $\frac{75\%}{90\%} = \frac{25\%}{x}$

To solve for x, the fractions must be cross multiplied: ($75\%x = 90\% \times 25\%$). To make things easier, let's convert the percentages to decimals: ($0.9 \times 0.25 = 0.225 = 0.75x$). To get rid of x's co-efficient, each side must be divided by that same coefficient to get the answer $x = 0.3$. The question could ask for the answer as a percentage or fraction in lowest terms, which are 30% and $\frac{3}{10}$, respectively.

2) $\frac{x}{12} = \frac{30}{96}$

Cross-multiply: $96x = 30 \times 12$
Multiply: $96x = 360$
Divide: $x = 360 \div 96$
Answer: $x = 3.75$

3) $\frac{0.5}{3} = \frac{x}{6}$

Cross-multiply: $3x = 0.5 \times 6$
Multiply: $3x = 3$
Divide: $x = 3 \div 3$
Answer: $x = 1$

You may have noticed there's a faster way to arrive at the answer. If there is an obvious operation being performed on the proportion, the same operation can be used on the other side of the proportion to solve for x. For example, in the first practice problem, 75% became 25% when divided by 3, and upon doing the same to 90%, the correct answer of 30% would have been found with much less legwork. However, these questions aren't always so intuitive, so it's a good idea to work through the steps, even if the answer seems apparent from the outset.

There is a specific order in which operations are performed:

Parentheses – calculate anything inside parentheses first.

Exponents – apply any exponents second.

Multiplication – execute any multiplication.

Division – execute any division.

Addition – execute any addition.

Subtraction – execute any subtraction.

A memory device to help recall the order is word PEMDAS, or "Please Excuse My Dear Aunt Sally."

Example
Solve using correct order of operations $(3 + 4)(4 \div 2) + 8$.

Calculate anything inside parentheses first.

$$(3 + 4)(4 \div 2) + 8$$

Multiply.

$$(7)(2) + 8$$

Add, then solve.

$$14 + 8 = 22$$

Not every equation contains every operator, but the order of operations needs to be followed to obtain the correct answer. Some tell you the number that represents a variable. In that case, replace the variable with the number first, then follow the order to solve.

Example

Solve $X^2 + 5 - 1$, for $X = 3$.

Replace X with 3.

$$X^2 + 5 - 1$$

Use order of operations to solve exponents.

$$3^2 + 5 - 1$$

Add and subtract.

$$9 + 5 - 1$$

Solve.

$$14 - 1 = 13$$

Another application for algebra is temperature conversion. The United States commonly uses the Fahrenheit scale to measure temperature. In science and medicine, the Celsius scale is used. Here are some common comparisons:

$$0\,^0C = 32\,^0F, \text{water freezes.}$$

$$100\,^0C = 212\,^0F, \text{water boils.}$$

To convert between the two temperature scales, use the following equations.

$$9/5\,^0C + 32 = {}^0F$$

$$5/9\,(^0F - 32) = {}^0C$$

Example

A patient has a temperature of 39.6 °C. Convert this to Fahrenheit to assess whether she needs medical care.

Set up the equation using 39.6 for °C.

$$9/5\,(39.6) + 32 = {}^0F$$

Multiply, divide and then add.

$$71.28 + 32 = {}^0F$$

Solve.

$$103.28 \,^{0}F$$

With a temperature over 103 °F, the patient may need medical care.

When trying to isolate a term or solve for a variable on one side of an equation, it is important not to change the equation. Always do the same operations to both sides of the equation.

Example
Solve for X, $X - 9 = 10$.

Solve for X by isolating it on a side.

$$X - 9 = 10$$

To get X alone, eliminate the 9 by adding 9 to both sides.

$$X - 9 + 9 = 10 + 9$$

Solve.

$$X = 19$$

Example
Solve for X, $4X = 20$.

Solve for X by isolating it on one side.

$$4x = 20$$

To get X alone, eliminate the 4 by dividing both sides by 4.

$$\frac{4x}{4} = \frac{20}{4}$$

Solve

$$X = 5$$

Example
Solve for x, $x^2 - 2 = 7$.

Isolate x on a side by adding 2 to both sides.

$$x^2 - 2 + 2 = 7 + 2$$

To undo the squaring of x, take the square root of both sides.

$$\sqrt{x^2} = \sqrt{9}$$

Solve.

$$x = 3$$

While solving an equation, you can also combine like terms. This is also called simplifying.

Examples of like terms would be x² and 3x², or 4x and 8x.

Simplify: $x^2 + 2x^2 + 9x - 3x + 1 - 5$

This is not a full equation so we cannot solve it, only simplify it.

Identify all like terms.

$$x^2 + 2x^2 + 9x - 3x + 1 - 5$$

Combine the terms. Be sure to use the proper signs.

$$3x^2 + 6x - 4$$

Ratio Problems

A *ratio* compares the size of one group to the size of another. For example, there may be a room with 4 tables and 24 chairs. The ratio of tables to chairs is 4: 24. Such ratios behave like fractions in that both sides of the ratio by the same number can be multiplied or divided. Thus, the ratio 4:24 is the same as the ratio 2:12 and 1:6.

One quantity is *proportional* to another quantity if the first quantity is always some multiple of the second. For instance, the distance travelled in five hours is always five times to the speed as travelled. The distance is proportional to speed in this case.

One quantity is *inversely proportional* to another quantity if the first quantity is equal to some number divided by the second quantity. The time it takes to travel one hundred miles will be given by 100 divided by the speed travelled. The time is inversely proportional to the speed.

When dealing with word problems, there is no fixed series of steps to follow, but there are some general guidelines to use. It is important that the quantity to be found is identified. Then, it can be determined how the given values can be used and manipulated to find the final answer.

Example 1

Jana wants to travel to visit Alice, who lives one hundred and fifty miles away. If she can drive at fifty miles per hour, how long will her trip take?

The quantity to find is the *time* of the trip. The time of a trip is given by the distance to travel divided by the speed to be traveled. The problem determines that the distance is one hundred and fifty miles, while the speed is fifty miles per hour. Thus, 150 divided by 50 is $150 \div 50 = 3$. Because *miles* and *miles per hour* are the units being divided, the miles cancel out. The result is 3 hours.

Example 2

Bernard wishes to paint a wall that measures twenty feet wide by eight feet high. It costs ten cents to paint one square foot. How much money will Bernard need for paint?

The final quantity to compute is the *cost* to paint the wall. This will be ten cents ($0.10) for each square foot of area needed to paint. The area to be painted is unknown, but the dimensions of the wall are given; thus, it can be calculated.

The dimensions of the wall are 20 feet wide and 8 feet high. Since the area of a rectangle is length multiplied by width, the area of the wall is 8 x 20 = 160 square feet. Multiplying 0.1 x 160 yields $16 as the cost of the paint.

The *average* or *mean* of a collection of numbers is given by adding those numbers together and then dividing by the total number of values. A *weighted average* or *weighted mean* is given by adding the numbers multiplied by their weights, then dividing by the sum of the weights:

$$\frac{w_1x_1 + w_2x_2 + w_3x_3 \dots + w_nx_n}{w_1 + w_2 + w_3 + \dots + w_n}$$

An *ordinary average* is a weighted average where all the weights are 1.

Unit Rate

Unit rate word problems will ask to calculate the rate or quantity of something in a different value. For example, a problem might say that a car drove a certain number of miles in a certain number of minutes and then ask how many miles per hour the car was traveling. These questions involve solving proportions. Consider the following examples:

1) Alexandra made $96 during the first 3 hours of her shift as a temporary worker at a law office. She will continue to earn money at this rate until she finishes in 5 more hours. How much does Alexandra make per hour? How much will Alexandra have made at the end of the day?

This problem can be solved in two ways. The first is to set up a proportion, as the rate of pay is constant. The second is to determine her hourly rate, multiply the 5 hours by that rate, and then add the $96.

To set up a proportion, put the money already earned over the hours already worked on one side of an equation. The other side has x over 8 hours (the total hours worked in the day). It looks like this: $\frac{96}{3} = \frac{x}{8}$. Now, cross-multiply to get $768 = 3x$. To get x, divide by 3, which leaves $x = 256$. Alternatively, as x is the numerator of one of the proportions, multiplying by its denominator will reduce the solution by one step. Thus, Alexandra will make $256 at the end of the day. To calculate her hourly rate, divide the total by 8, giving $32 per hour.

Alternatively, it is possible to figure out the hourly rate by dividing $96 by 3 hours to get $32 per hour. Now her total pay can be figured by multiplying $32 per hour by 8 hours, which comes out to $256.

2) Jonathan is reading a novel. So far, he has read 215 of the 335 total pages. It takes Jonathan 25 minutes to read 10 pages, and the rate is constant. How long does it take Jonathan to read one page? How much longer will it take him to finish the novel? Express the answer in time.

To calculate how long it takes Jonathan to read one page, divide the 25 minutes by 10 pages to determine the page per minute rate. Thus, it takes 2.5 minutes to read one page.

Jonathan must read 120 more pages to complete the novel. (This is calculated by subtracting the pages already read from the total.) Now, multiply his rate per page by the number of pages. Thus, $120 \div 2.5 = 300$. Expressed in time, 300 minutes is equal to 5 hours.

3) At a hotel, $\frac{4}{5}$ of the 120 rooms are booked for Saturday. On Sunday, $\frac{3}{4}$ of the rooms are booked. On which day are more of the rooms booked, and by how many more?

The first step is to calculate the number of rooms booked for each day. Do this by multiplying the fraction of the rooms booked by the total number of rooms.

Saturday: $\frac{4}{5} \times 120 = \frac{4}{5} \times \frac{120}{1} = \frac{480}{5} = 96$ rooms

Sunday: $\frac{3}{4} \times 120 = \frac{3}{4} \times \frac{120}{1} = \frac{360}{4} = 90$ rooms

Thus, more rooms were booked on Saturday by 6 rooms.

4) In a veterinary hospital, the veterinarian-to-pet ratio is 1:9. The ratio is always constant. If there are 45 pets in the hospital, how many veterinarians are currently in the veterinary hospital?

Set up a proportion to solve for the number of veterinarians: $\frac{1}{9} = \frac{x}{45}$

Cross-multiplying results in $9x = 45$, which works out to 5 veterinarians.

Alternatively, as there are always 9 times as many pets as veterinarians, is it possible to divide the number of pets (45) by 9. This also arrives at the correct answer of 5 veterinarians.

5) At a general practice law firm, 30% of the lawyers work solely on tort cases. If 9 lawyers work solely on tort cases, how many lawyers work at the firm?

First, solve for the total number of lawyers working at the firm, which will be represented here with x. The problem states that 9 lawyers work solely on torts cases, and they make up 30% of the total lawyers at the firm. Thus, 30% multiplied by the total, x, will equal 9. Written as equation, this is: $30\% \times x = 9$.

It's easier to deal with the equation after converting the percentage to a decimal, leaving $0.3x = 9$. Thus, $x = \frac{9}{0.3} = 30$ lawyers working at the firm.

6) Xavier was hospitalized with pneumonia. He was originally given 35mg of antibiotics. Later, after his condition continued to worsen, Xavier's dosage was increased to 60mg. What was the percent increase of the antibiotics? Round the percentage to the nearest tenth.

An increase or decrease in percentage can be calculated by dividing the difference in amounts by the original amount and multiplying by 100. Written as an equation, the formula is:

$$\frac{new\ quantity - old\ quantity}{old\ quantity} \times 100$$

Here, the question states that the dosage was increased from 35mg to 60mg, so these are plugged into the formula to find the percentage increase.

$$\frac{60 - 35}{35} \times 100 = \frac{25}{35} \times 100$$

$$0.7142 \times 100 = 71.4\%$$

Rate of Change

Rate of change for any line calculates the steepness of the line over a given interval. Rate of change is also known as the slope or rise/run. The TEAS will focus on the rate of change for linear functions which are straight lines. The slope is given by the change in *y* divided by the change in *x*. So the formula looks like this:

$$slope = \frac{y_2 - y_1}{x_2 - x_1}$$

In the graph below, two points are plotted. The first has the coordinates of (0,1) and the second point is (2,3). Remember that the x coordinate is always placed first in coordinate pairs. Work from left to right when identifying coordinates. Thus, the point on the left is point 1 (0,1) and the point on the right is point 2 (2,3).

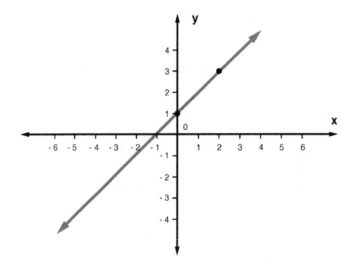

Now we need to just plug those numbers into the equation:

$$slope = \frac{3 - 1}{2 - 0}$$

$$slope = \frac{2}{2}$$

$$slope = 1$$

This means that for every increase of 1 for x, y also increased by 1. You can see this in the line. When x equalled 0, y equalled 1, and when x was increased to 1, y equalled 2.

Slope can be thought of as determining the rise over run:

$$slope = \frac{rise}{run}$$

The rise being the change vertically on the y axis and the run being the change horizontally on the x axis.

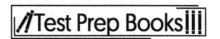

Converting Within and Between Standard and Metric Systems

American Measuring System

The measuring system used today in the United States developed from the British units of measurement during colonial times. The most typically used units in this customary system are those used to measure weight, liquid volume, and length, whose common units are found below. In the customary system, the basic unit for measuring weight is the ounce (oz); there are 16 ounces (oz) in 1 pound (lb) and 2000 pounds in 1 ton. The basic unit for measuring liquid volume is the ounce (oz); 1 ounce is equal to 2 tablespoons (tbsp) or 6 teaspoons (tsp), and there are 8 ounces in 1 cup, 2 cups in 1 pint (pt), 2 pints in 1 quart (qt), and 4 quarts in 1 gallon (gal). For measurements of length, the inch (in) is the base unit; 12 inches make up 1 foot (ft), 3 feet make up 1 yard (yd), and 5280 feet make up 1 mile (mi). However, as there are only a set number of units in the customary system, with extremely large or extremely small amounts of material, the numbers can become awkward and difficult to compare.

Common Customary Measurements		
Length	Weight	Capacity
1 foot = 12 inches	1 pound = 16 ounces	1 cup = 8 fluid ounces
1 yard = 3 feet	1 ton = 2,000 pounds	1 pint = 2 cups
1 yard = 36 inches		1 quart = 2 pints
1 mile = 1,760 yards		1 quart = 4 cups
1 mile = 5,280 feet		1 gallon = 4 quarts
		1 gallon = 16 cups

Metric System

Aside from the United States, most countries in the world have adopted the metric system embodied in the International System of Units (SI). The three main SI base units used in the metric system are the meter (m), the kilogram (kg), and the liter (L); meters measure length, kilograms measure mass, and liters measure volume.

These three units can use different prefixes, which indicate larger or smaller versions of the unit by powers of ten. This can be thought of as making a new unit which is sized by multiplying the original unit in size by a factor.

These prefixes and associated factors are:

Metric Prefixes			
Prefix	Symbol	Multiplier	Exponential
kilo	k	1,000	10^3
hecto	h	100	10^2
deca	da	10	10^1
no prefix		1	10^0
deci	d	0.1	10^{-1}
centi	c	0.01	10^{-2}
milli	m	0.001	10^{-3}

The correct prefix is then attached to the base. Some examples:

1 milliliter equals .001 liters.

1 kilogram equals 1,000 grams.

Choosing the Appropriate Measuring Unit

Some units of measure are represented as square or cubic units depending on the solution. For example, perimeter is measured in units, area is measured in square units, and volume is measured in cubic units.

Also be sure to use the most appropriate unit for the thing being measured. A building's height might be measured in feet or meters while the length of a nail might be measured in inches or centimeters. Additionally, for SI units, the prefix should be chosen to provide the most succinct available value. For example, the mass of a bag of fruit would likely be measured in kilograms rather than grams or milligrams, and the length of a bacteria cell would likely be measured in micrometers rather than centimeters or kilometers.

Conversion

Converting measurements in different units between the two systems can be difficult because they follow different rules. The best method is to look up an English to Metric system conversion factor and then use a series of equivalent fractions to set up an equation to convert the units of one of the measurements into those of the other.

The table below lists some common conversion values that are useful for problems involving measurements with units in both systems:

English System	Metric System
1 inch	2.54 cm
1 foot	0.3048 m
1 yard	0.914 m
1 mile	1.609 km
1 ounce	28.35 g
1 pound	0.454 kg
1 fluid ounce	29.574 mL
1 quart	0.946 L
1 gallon	3.785 L

Consider the example where a scientist wants to convert 6.8 inches to centimeters. The table above is used to find that there are 2.54 centimeters in every inch, so the following equation should be set up and solved:

$$\frac{6.8\ in}{1} \times \frac{2.54\ cm}{1\ in} = 17.272\ cm$$

Notice how the inches in the numerator of the initial figure and the denominator of the conversion factor cancel out. (This equation could have been written simply as $6.8\ in \times 2.54\ cm = 17.272\ cm$, but it was shown in detail to illustrate the steps). The goal in any conversion equation is to set up the fractions so that the units you are trying to convert from cancel out and the units you desire remain.

For a more complicated example, consider converting 2.15 kilograms into ounces. The first step is to convert kilograms into grams and then grams into ounces. Note that the measurement you begin with does not have to be put in a fraction.

So in this case, 2.15 kg is by itself although it's technically the numerator of a fraction:

$$2.15 \; kg \; \times \frac{1000g}{kg} = 2150 \; g$$

Then, use the conversion factor from the table to convert grams to ounces:

$$2150g \; \times \frac{1 \; oz}{28.35g} = 75.8 \; oz$$

Choose a Level of Accuracy Appropriate to Limitations on Measurements

Precision and Accuracy

Precision and accuracy are used to describe groups of measurements. *Precision* describes a group of measures that are very close together, regardless of whether the measures are close to the true value. *Accuracy* describes how close the measures are to the true value.

Since accuracy refers to the closeness of a value to the true measurement, the level of accuracy depends on the object measured and the instrument used to measure it. This will vary depending on the situation. If measuring the mass of a set of dictionaries, kilograms may be used as the units. In this case, it is not vitally important to have a high level of accuracy. If the measurement is a few grams away from the true value, the discrepancy might not make a big difference in the problem.

In a different situation, the level of accuracy may be more significant. Pharmacists need to be sure they are very accurate in their measurements of medicines that they give to patients. In this case, the level of accuracy is vitally important and not something to be estimated. In the dictionary situation, the measurements were given as whole numbers in kilograms. In the pharmacist's situation, the measurements for medicine must be taken to the milligram and sometimes further, depending on the type of medicine.

When considering the accuracy of measurements, the error in each measurement can be shown as absolute and relative. *Absolute error* tells the actual difference between the measured value and the true value. The *relative error* tells how large the error is in relation to the true value. There may be two problems where the absolute error of the measurements is 10 grams. For one problem, this may mean the relative error is very small because the measured value is 14,990 grams, and the true value is 15,000 grams. Ten grams in relation to the true value of 15,000 is small: 0.06%. For the other problem, the measured value is 290 grams, and the true value is 300 grams. In this case, the 10-gram absolute error means a high relative error because the true value is smaller. The relative error is 10/300 = 0.03, or 3%.

Estimation

Estimation is finding a value that is close to a solution but is not the exact answer. For example, if there are values in the thousands to be multiplied, then each value can be estimated to the nearest thousand

and the calculation performed. This value provides an approximate solution that can be determined very quickly.

Rounding Numbers

It's often convenient to round a number, which means to give an approximate figure to make it easier to compare amounts or perform mental math. Round up when the digit is 5 or more. The rounded The digit used to determine the rounding, and all subsequent digits, become 0, and the selected place value is increased by 1. Here are some examples:

75 rounded to the nearest ten is 80
380 rounded to the nearest hundred is 400
22.697 rounded to the nearest hundredth is 22.70

Round down when rounding on any digit that is below 5. The rounded digit, and all subsequent digits, becomes 0, and the preceding digit goes down by 1. Here are some examples:

92 rounded to the nearest ten is 90
839 rounded to the nearest hundred is 800
22.643 rounded to the nearest hundredth is 22.64

Determining the Reasonableness of Results

When solving math word problems, the solution obtained should make sense within the given scenario. The step of checking the solution will reduce the possibility of a calculation error or a solution that may be *mathematically* correct but not applicable in the real world. Consider the following scenarios:

A problem states that Lisa got 24 out of 32 questions correct on a test and asks to find the percentage of correct answers. To solve the problem, a student divided 32 by 24 to get 1.33, and then multiplied by 100 to get 133 percent. By examining the solution within the context of the problem, the student should recognize that getting all 32 questions correct will produce a perfect score of 100 percent. Therefore, a score of 133 percent with 8 incorrect answers does not make sense and the calculations should be checked.

A problem states that the maximum weight on a bridge cannot exceed 22,000 pounds. The problem asks to find the maximum number of cars that can be on the bridge at one time if each car weighs 4,000 pounds. To solve this problem, a student divided 22,000 by 4,000 to get an answer of 5.5. By examining the solution within the context of the problem, the student should recognize that although the calculations are mathematically correct, the solution does not make sense. Half of a car on a bridge is not possible, so the student should determine that a maximum of 5 cars can be on the bridge at the same time.

Mental Math Estimation

Once a result is determined to be logical within the context of a given problem, the result should be evaluated by its nearness to the expected answer. This is performed by approximating given values to perform mental math. Numbers should be rounded to the nearest value possible to check the initial results.

Consider the following example: A problem states that a customer is buying a new sound system for their home. The customer purchases a stereo for $435, 2 speakers for $67 each, and the necessary

cables for $12. The customer chooses an option that allows him to spread the costs over equal payments for 4 months. How much will the monthly payments be?

After making calculations for the problem, a student determines that the monthly payment will be $145.25. To check the accuracy of the results, the student rounds each cost to the nearest ten ($440 + 70 + 70 + 10$) and determines that the total is approximately $590. Dividing by 4 months gives an approximate monthly payment of $147.50. Therefore, the student can conclude that the solution of $145.25 is very close to what should be expected.

When rounding, the place-value that is used in rounding can make a difference. Suppose the student had rounded to the nearest hundred for the estimation. The result ($400 + 100 + 100 + 0 = 600$; $600 \div 4 = 150$) will show that the answer is reasonable, but not as close to the actual value as rounding to the nearest ten.

Algebra

Interpreting Parts of an Expression

A fraction, or ratio, wherein each part is a polynomial, defines *rational expressions*. Some examples include $\frac{2x+6}{x}$, $\frac{1}{x^2-4x+8}$, and $\frac{z^2}{x+5}$. Exponents on the variables are restricted to whole numbers, which means roots and negative exponents are not included in rational expressions.

Rational expressions can be transformed by factoring. For example, the expression $\frac{x^2-5x+6}{(x-3)}$ can be rewritten by factoring the numerator to obtain:

$$\frac{(x-3)(x-2)}{(x-3)}$$

Therefore, the common binomial $(x-3)$ can cancel so that the simplified expression is:

$$\frac{(x-2)}{1} = (x-2)$$

Additionally, other rational expressions can be rewritten to take on different forms. Some may be factorable in themselves, while others can be transformed through arithmetic operations. Rational expressions are closed under addition, subtraction, multiplication, and division by a nonzero expression. *Closed* means that if any one of these operations is performed on a rational expression, the result will still be a rational expression. The set of all real numbers is another example of a set closed under all four operations.

Adding and subtracting rational expressions is based on the same concepts as adding and subtracting simple fractions. For both concepts, the denominators must be the same for the operation to take place. For example, here are two rational expressions:

$$\frac{x^3-4}{(x-3)} + \frac{x+8}{(x-3)}$$

Since the denominators are both $(x - 3)$, the numerators can be combined by collecting like terms to form:

$$\frac{x^3 + x + 4}{(x - 3)}$$

If the denominators are different, they need to be made common (the same) by using the Least Common Denominator (LCD). Each denominator needs to be factored, and the LCD contains each factor that appears in any one denominator the greatest number of times it appears in any denominator. The original expressions need to be multiplied times a form of 1, which will turn each denominator into the LCD. This process is like adding fractions with unlike denominators. It is also important when working with rational expressions to define what value of the variable makes the denominator zero. For this particular value, the expression is undefined.

Multiplication of rational expressions is performed like multiplication of fractions. The numerators are multiplied; then, the denominators are multiplied. The final fraction is then simplified. The expressions are simplified by factoring and cancelling out common terms. In the following example, $\frac{x^2}{(x-4)} * \frac{x^2-x-12}{2}$, the numerator of the second expression can be factored first to simplify the expression before multiplying. It turns into $\frac{x^2}{(x-4)} * \frac{(x-4)(x+3)}{2}$, where the $(x - 4)$ cancels out on the top and bottom, leaving:

$$\frac{x^2}{1} * \frac{(x + 3)}{2}$$

Then multiplication is performed, resulting in:

$$\frac{x^3 + 3x^2}{2}$$

Dividing rational expressions is similar to the division of fractions, where division turns into multiplying by a reciprocal. Therefore, given $\frac{x^2-3x+7}{x-4} \div \frac{x^2-5x+3}{x-4}$, the expression is rewritten as a multiplication problem:

$$\frac{x^2 - 3x + 7}{x - 4} * \frac{x - 4}{x^2 - 5x + 3}$$

The $x - 4$ cancels out, leaving:

$$\frac{x^2 - 3x + 7}{x^2 - 5x + 3}$$

The final answers should always be completely simplified. If a function is composed of a rational expression, the zeros of the graph can be found from setting the polynomial in the numerator as equal to zero and solving. The values that make the denominator equal to zero will either exist on the graph as a hole or a vertical asymptote.

Use the Structure of an Expression to Identify Ways to Rewrite It

Algebraic expressions are made up of numbers, variables, and combinations of the two, using mathematical operations. Expressions can be rewritten based on their factors. For example, the expression $6x + 4$ can be rewritten as $2(3x + 2)$ because 2 is a factor of both $6x$ and 4. More complex expressions can also be rewritten based on their factors. The expression $x^4 - 16$ can be rewritten as $(x^2 - 4)(x^2 + 4)$. This is a different type of factoring, where a difference of squares is factored into a sum and difference of the same two terms. With some expressions, the factoring process is simple and only leads to a different way to represent the expression. With others, factoring and rewriting the expression leads to more information about the given problem.

In the following quadratic equation, factoring the binomial leads to finding the zeros of the function: $x^2 - 5x + 6 = y$. This equations factors into $(x - 3)(x - 2) = y$, where 2 and 3 are found to be the zeros of the function when y is set equal to zero. The zeros of any function are the x-values where the graph of the function on the coordinate plane crosses the x-axis.

Factoring an equation is a simple way to rewrite the equation and find the zeros, but factoring is not possible for every quadratic. Completing the square is one way to find zeros when factoring is not an option. The following equation cannot be factored:

$$x^2 + 10x - 9 = 0$$

The first step in this method is to move the constant to the right side of the equation, making it:

$$x^2 + 10x = 9$$

Then, the coefficient of x is divided by 2 and squared. This number is then added to both sides of the equation, to make the equation still true. For this example, $\left(\frac{10}{2}\right)^2 = 25$ is added to both sides of the equation to obtain:

$$x^2 + 10x + 25 = 9 + 25$$

This expression simplifies to $x^2 + 10x + 25 = 34$, which can then be factored into:

$$(x + 5)^2 = 34$$

Solving for x then involves taking the square root of both sides and subtracting 5. This leads to two zeros of the function:

$$x = \pm\sqrt{34} - 5$$

Depending on the type of answer the question seeks, a calculator may be used to find exact numbers.

Given a quadratic equation in standard form— $ax^2 + bx + c = 0$ —the sign of a tells whether the function has a minimum value or a maximum value. If $a > 0$, the graph opens up and has a minimum value. If $a < 0$, the graph opens down and has a maximum value. Depending on the way the quadratic equation is written, multiplication may need to occur before a max/min value is determined.

Exponential expressions can also be rewritten, just as quadratic equations. Properties of exponents must be understood. Multiplying two exponential expressions with the same base involves adding the exponents:

$$a^m a^n = a^{m+n}$$

Dividing two exponential expressions with the same base involves subtracting the exponents:

$$\frac{a^m}{a^n} = a^{m-n}$$

Raising an exponential expression to another exponent includes multiplying the exponents:

$$(a^m)^n = a^{mn}$$

The zero power always gives a value of 1:

$$a^0 = 1$$

Raising either a product or a fraction to a power involves distributing that power:

$$(ab)^m = a^m b^m \text{ and } \left(\frac{a}{b}\right)^m = \frac{a^m}{b^m}$$

Finally, raising a number to a negative exponent is equivalent to the reciprocal including the positive exponent:

$$a^{-m} = \frac{1}{a^m}$$

Factor a Quadratic Functions to Reveal the Zeros

A polynomial of degree 2 is called *quadratic*. Every quadratic function can be written in the form:

$$ax^2 + bx + c$$

The graph of a quadratic function, $y = ax^2 + bx + c$, is called a *parabola*. Parabolas are vaguely U-shaped.

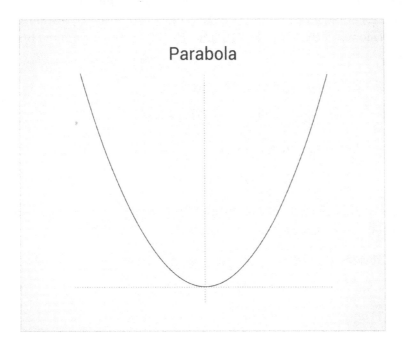

Parabola

Whether the parabola opens upward or downward depends on the sign of *a*. If *a* is positive, then the parabola will open upward. If *a* is negative, then the parabola will open downward. The value of *a* will also affect how wide the parabola is. If the absolute value of *a* is large, then the parabola will be fairly skinny. If the absolute value of *a* is small, then the parabola will be quite wide.

Changes to the value of *b* affect the parabola in different ways, depending on the sign of *a*. For positive values of *a*, increasing *b* will move the parabola to the left, and decreasing *b* will move the parabola to the right. On the other hand, if *a* is negative, the effects will be the opposite: increasing *b* will move the parabola to the right, while decreasing *b* will move the parabola to the left.

Changes to the value of *c* move the parabola vertically. The larger that *c* is, the higher the parabola gets. This does not depend on the value of *a*.

The quantity $D = b^2 - 4ac$ is called the *discriminant* of the parabola. When the discriminant is positive, then the parabola has two real zeros, or *x* intercepts. However, if the discriminant is negative, then there are no real zeros, and the parabola will not cross the *x*-axis. The highest or lowest point of the parabola is called the *vertex*. If the discriminant is zero, then the parabola's highest or lowest point is on the *x*-axis, and it will have a single real zero. The x-coordinate of the vertex can be found using the equation:

$$x = -\frac{b}{2a}$$

Plug this x-value into the equation and find the y-coordinate.

A quadratic equation is often used to model the path of an object thrown into the air. The x-value can represent the time in the air, while the y-value can represent the height of the object. In this case, the maximum height of the object would be the y-value found when the x-value is $-\frac{b}{2a}$.

Factoring

Factors for polynomials are similar to factors for integers—they are numbers, variables, or polynomials that, when multiplied together, give a product equal to the polynomial in question. One polynomial is a factor of a second polynomial if the second polynomial can be obtained from the first by multiplying by a third polynomial.

$6x^6 + 13x^4 + 6x^2$ can be obtained by multiplying together:

$$(3x^4 + 2x^2)(2x^2 + 3)$$

This means $2x^2 + 3$ and $3x^4 + 2x^2$ are factors of:

$$6x^6 + 13x^4 + 6x^2$$

In general, finding the factors of a polynomial can be tricky. However, there are a few types of polynomials that can be factored in a straightforward way.

If a certain monomial divides each term of a polynomial, it can be factored out:

$$x^2 + 2xy + y^2 = (x + y)^2$$
$$x^2 - 2xy + y^2 = (x - y)^2$$
$$x^2 - y^2 = (x + y)(x - y)$$
$$x^3 + y^3 = (x + y)(x^2 - xy + y^2)$$
$$x^3 - y^3 = (x - y)(x^2 + xy + y^2)$$
$$x^3 + 3x^2y + 3xy^2 + y^3 = (x + y)^3$$
$$x^3 - 3x^2y + 3xy^2 - y^3 = (x - y)^3$$

These rules can be used in many combinations with one another. For example, the expression $3x^3 - 24$ factors to:

$$3(x^3 - 8) = 3(x - 2)(x^2 + 2x + 4)$$

When factoring polynomials, a good strategy is to multiply the factors to check the result.

Converting Standard Form to Factored Form

To factor an expression, a greatest common factor needs to be factored out first. Then, if possible, the remaining expression needs to be factored into the product of binomials. A binomial is an expression with two terms.

Greatest Common Factor

The greatest common factor (GCF) of a monomial (one term) consists of the largest number that divides evenly into all coefficients (number part of a term); and if all terms contain the same variable, the variable with the lowest exponent. The GCF of $3x^4 - 9x^3 + 12x^2$ would be $3x^2$. To write the factored expression, every term needs to be divided by the GCF, then the product of the resulting quotient and the GCF (using parentheses to show multiplication) should be written. For the previous example, the factored expression would be:

$$3x^2(x^2 - 3x + 4)$$

Factoring Ax² + Bx + C When A = 1

To factor a quadratic expression in standard form when the value of a is equal to 1, the factors that multiply to equal the value of c should be found and then added to equal the value of b (the signs of b and c should be included). The factored form for the expression will be the product of binomials:

$$(x + factor1)(x + factor2)$$

Here's a sample expression:

$$x^2 - 4x - 5$$

The two factors that multiply to equal $c(-5)$ and add together to equal $b(-4)$ are -5 and 1. Therefore, the factored expression would be:

$$(x - 5)(x + 1)$$

Note $(x + 1)(x - 5)$ is equivalent.

Factoring a Difference of Squares

A difference of squares is a binomial expression where both terms are perfect squares (perfect square-perfect square). Perfect squares include 1, 4, 9, 16 . . . and x^2, x^4, x^6 . . .

The factored form of a difference of squares will be:

$$(\sqrt{term1} + \sqrt{term2})(\sqrt{term1} - \sqrt{term2})$$

For example:

$$x^2 - 4 = (x + 2)(x - 2)$$

And

$$25x^6 - 81 = (5x^3 + 9)(5x^3 - 9)$$

Factoring *Ax² + Bx + C when A ≠ 1*

To factor a quadratic expression in standard form when the value of a is not equal to 1, the factors that multiply to equal the value of $a \times c$ should be found and then added to equal the value of b. Next, the expression splitting the bx term should be rewritten using those factors. Instead of three terms, there will now be four. Then the first two terms should be factored using GCF, and a common binomial should be factored from the last two terms. The factored form will be: (common binomial) (2 terms out of binomials). In the sample expression $2 \times 2 + 11x + 12$, the value of:

$$a \times c \ (2 \times 12) = 24$$

Two factors that multiply to 24 and added together to yield $b(11)$ are 8 and 3. The bx term ($11x$) can be rewritten by splitting it into the factors:

$$2 \times 2 + 8x + 3x + 12$$

A GCF from the first two terms can be factored as:

$$2x(x + 4) + 3x + 12$$

A common binomial from the last two terms can then be factored as:

$$2(x + 4) + 3(x + 4)$$

The factored form can be written as a product of binomials: $(x + 4)(2x + 3)$.

Converting Factored Form to Standard Form

To convert an expression from factored form to standard form, the factors are multiplied.

Zeros of Polynomials

Finding the zeros of polynomial functions is the same process as finding the solutions of polynomial equations. These are the points at which the graph of the function crosses the x-axis. As stated previously, factors can be used to find the zeros of a polynomial function. The degree of the function shows the number of possible zeros. If the highest exponent on the independent variable is 4, then the degree is 4, and the number of possible zeros is 4. If there are complex solutions, the number of roots is less than the degree.

Given the function $y = x^2 + 7x + 6$, y can be set equal to zero, and the polynomial can be factored. The equation turns into $0 = (x + 1)(x + 6)$, where $x = -1$ and $x = -6$ are the zeros. Since this is a quadratic equation, the shape of the graph will be a parabola. Knowing that zeros represent the points where the parabola crosses the x-axis, the maximum or minimum point is the only other piece needed to sketch a rough graph of the function. By looking at the function in standard form, the coefficient of x is positive; therefore, the parabola opens *up*.

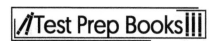

Using the zeros and the minimum, the following rough sketch of the graph can be constructed:

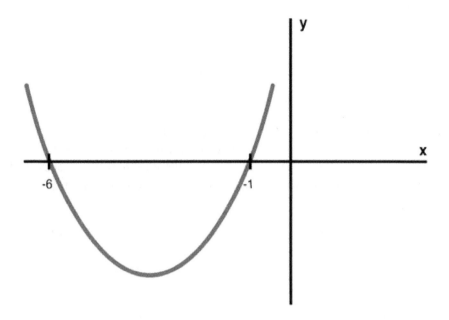

Understand that Polynomials Form a System Analogous to the Integers

An expression of the form ax^n, where n is a non-negative integer, is called a *monomial* because it contains one term. A sum of monomials is called a *polynomial*. For example, $-4x^3 + x$ is a polynomial, while $5x^7$ is a monomial. A function equal to a polynomial is called a *polynomial function*.

The monomials in a polynomial are also called the *terms* of the polynomial.

The constants that precede the variables are called *coefficients*.

The highest value of the exponent of x in a polynomial is called the *degree* of the polynomial. So, $-4x^3 + x$ has a degree of 3, while $-2x^5 + x^3 + 4x + 1$ has a degree of 5. When multiplying polynomials, the degree of the result will be the sum of the degrees of the two polynomials being multiplied.

To add polynomials, add the coefficients of like powers of x. For example:

$$(-2x^5 + x^3 + 4x + 1) + (-4x^3 + x)$$

$$-2x^5 + (1 - 4)x^3 + (4 + 1)x + 1$$

$$-2x^5 - 3x^3 + 5x + 1$$

Likewise, subtraction of polynomials is performed by subtracting coefficients of like powers of x. So,

$$(-2x^5 + x^3 + 4x + 1) - (-4x^3 + x)$$

$$-2x^5 + (1 + 4)x^3 + (4 - 1)x + 1$$

$$-2x^5 + 5x^3 + 3x + 1$$

To multiply two polynomials, multiply each term of the first polynomial by each term of the second polynomial and add the results. For example:

$$(4x^2 + x)(-x^3 + x)$$

$$4x^2(-x^3) + 4x^2(x) + x(-x^3) + x(x)$$

$$-4x^5 + 4x^3 - x^4 + x^2$$

In the case where each polynomial has two terms, like in this example, some students find it helpful to remember this as multiplying the First terms, then the Outer terms, then the Inner terms, and finally the Last terms, with the mnemonic FOIL. For longer polynomials, the multiplication process is the same, but there will be, of course, more terms, and there is no common mnemonic to remember each combination.

The process of *factoring* a polynomial means to write the polynomial as a product of other (generally simpler) polynomials. Here is an example:

$$x^2 - 4x + 3 = (x - 1)(x - 3)$$

If a certain monomial divides every term of the polynomial, factor it out of each term, for example:

$$4x^3 + 16x^2 = 4x^2(x + 4).$$

$$x^2 + 2xy + y^2 = (x + y)^2 \text{ or } x^2 - 2xy + y^2 = (x - y)^2$$

$$x^2 - y^2 = (x + y)(x - y)$$

$$x^3 + y^3 = (x + y)(x^2 - xy + y^2)$$

$$x^3 - y^3 = (x - y)(x^2 + xy + y^2)$$

$$x^3 + 3x^2y + 3xy^2 + y^3 = (x + y)^3 \text{ and } x^3 - 3x^2y + 3xy^2 - y^3 = (x - y)^3$$

It sometimes can be necessary to rewrite the polynomial in some clever way before applying the above rules. Consider the problem of factoring $x^4 - 1$. This does not immediately look like any of the cases for which there are rules. However, it's possible to think of this polynomial as $x^4 - 1 = (x^2)^2 - (1^2)^2$, and now apply the third rule in the above list to simplify this:

$$(x^2)^2 - (1^2)^2 = (x^2 + 1^2)(x^2 - 1^2)$$

$$(x^2 + 1)(x^2 - 1)$$

Adding and Subtracting Monomials and Polynomials

Addition and subtraction operations can be performed on polynomials with like terms. *Like terms refers to terms* that have the same variable and exponent. The two following polynomials can be added together by collecting like terms:

$$(x^2 + 3x - 4) + (4x^2 - 7x + 8)$$

The x^2 terms can be added as:

$$x^2 + 4x^2 = 5x^2$$

The x terms can be added as $3x + -7x = -4x$, and the constants can be added as:

$$-4 + 8 = 4$$

The following expression is the result of the addition:

$$5x^2 - 4x + 4$$

When subtracting polynomials, the same steps are followed, only subtracting like terms together.

Multiplication of polynomials can also be performed. Given the two polynomials, $(y^3 - 4)$ and $(x^2 + 8x - 7)$, each term in the first polynomial must be multiplied by each term in the second polynomial.

The steps to multiply each term in the given example are as follows:

$$(y^3 * x^2) + (y^3 * 8x) + (y^3 * -7) + (-4 * x^2) + (-4 * 8x) + (-4 * -7)$$

Simplifying each multiplied part, yields:

$$x^2 y^3 + 8xy^3 - 7y^3 - 4x^2 - 32x + 28$$

None of the terms can be combined because there are no like terms in the final expression. Any polynomials can be multiplied by each other by following the same set of steps, then collecting like terms at the end.

Multiplying and Dividing Monomials and Polynomials

To multiply polynomials each term of the first polynomial multiplies each term of the second polynomial, and adds up the results. Here's an example:

$$(3x^4 + 2x^2)(2x^2 + 3)$$

$$3x^4 \cdot 2x^2 + 3x^4 \cdot 3 + 2x^2 \cdot 2x^2 + 2x^2 \cdot 3$$

Then, add like terms with a result of:

$$6x^6 + 9x^4 + 4x^4 + 6x^2 = 6x^6 + 13x^4 + 6x^2$$

A polynomial with two terms is called a *binomial*. Another way of remember the rule for multiplying two binomials is to use the acronym *FOIL*: multiply the *First* terms together, then the *Outside* terms (terms on the far left and far right), then the *Inner* terms, and finally the *Last* two terms. For longer polynomials, there is no such convenient mnemonic, so remember to multiply each term of the first polynomial by each term of the second, and add the results.

To divide one polynomial by another, the procedure is similar to long division. At each step, one needs to figure out how to get the term of the dividend with the highest exponent as a multiple of the divisor. The divisor is multiplied by the multiple to get that term, which goes in the quotient. Then, the product of this term is subtracted with the dividend from the divisor and repeat the process. This sounds rather abstract, so it may be easiest to see the procedure by describing it while looking at an example.

Example
$(4x^3 + x^2 - x + 4) \div (2x - 1)$

The first step is to cancel out the highest term in the first polynomial.

To get $4x^3$ from the second polynomial, multiply by $2x^2$.

The first term for the quotient is going to be $2x^2$.

The result of $2x^2(2x - 1)$ is $4x^3 - 2x^2$. Subtract this from the first polynomial.

The result is $(-x^2 - x + 4) \div (2x - 1)$.

The procedure is repeated: to cancel the $-x^2$ term, then multiply $(2x - 1)$ by $-\frac{1}{2}x$.

Adding this to the quotient, the quotient becomes $2x^2 - \frac{1}{2}x$.

The dividend is changed by subtracting $-\frac{1}{2}x(2x - 1)$ from it to obtain $(-\frac{3}{2}x + 4) \div (2x - 1)$.

To get $-\frac{3}{2}x$ needs to be multiplied by $-\frac{3}{4}$.

The quotient, therefore, becomes $2x^2 - \frac{1}{2}x - \frac{3}{4}$.

The remaining part is $4.75 \div (2x - 1)$.

There is no monomial to multiply to cancel this constant term, since the divisor now has a higher power than the dividend.

The final answer is the quotient plus the remainder divided by $(2x - 1)$:

$$2x^2 - \frac{1}{2}x - \frac{3}{4} + \frac{4.75}{2x - 1}$$

Polynomial Identities

Difference of squares refers to a binomial composed of the difference of two squares. For example, $a^2 - b^2$ is a difference of squares. It can be written $(a)^2 - (b)^2$, and it can be factored into $(a - b)(a + b)$. Recognizing the difference of squares allows the expression to be rewritten easily because of the form it takes. For some expressions, factoring consists of more than one step. When factoring, it's important to always check to make sure that the result cannot be factored further. If it can, then the expression should be split further. If it cannot be, the factoring step is complete, and the expression is completely factored.

A sum and difference of cubes is another way to factor a polynomial expression. When the polynomial takes the form of addition or subtraction of two terms that can be written as a cube, a formula is given. The following graphic shows the factorization of a difference of cubes:

$$a^3 - b^3 = (a - b)(a^2 + ab + b^2)$$

same sign

opposite sign

always +

This form of factoring can be useful in finding the zeros of a function of degree 3. For example, when solving $x^3 - 27 = 0$, this rule needs to be used. $x^3 - 27$ is first written as the difference two cubes, $(x)^3 - (3)^3$ and then factored into:

$$(x - 3)(x^2 + 3x + 9)$$

This expression may not be factored any further. Each factor is then set equal to zero. Therefore, one solution is found to be $x = 3$, and the other two solutions must be found using the quadratic formula. A sum of squares would have a similar process. The formula for factoring a sum of squares is:

$$a^3 + b^3 = (a + b)(a^2 - ab + b^2)$$

The opposite of factoring is multiplying. Multiplying a square of a binomial involves the following rules:

$$(a + b)^2 = a^2 + 2ab + b^2$$

$$(a - b)^2 = a^2 - 2ab + b^2$$

The binomial theorem for expansion can be used when the exponent on a binomial is larger than 2, and the multiplication would take a long time. The binomial theorem is given as:

$$(a + b)^n = \sum_{k=0}^{n} \binom{n}{k} a^{n-k} b^k$$

$$\text{where} \quad \binom{n}{k} = \frac{n!}{k!(n-k)!}$$

The *Remainder Theorem* can be helpful when evaluating polynomial functions $P(x)$ for a given value of x. A polynomial can be divided by $(x - a)$, if there is a remainder of 0. This also means that $P(a) = 0$ and $(x - a)$ is a factor of $P(x)$. In a similar sense, if P is evaluated at any other number b, $P(b)$ is equal to the remainder of dividing $P(x)$ by $(x - b)$.

The Factor Theorem

Suppose $P(x)$ is a polynomial. The *Factor Theorem* states c, either a real or complex number, is a zero of the polynomial if, and only if, $(x - c)$ is a factor of $P(x)$. The proof involves first stating that from the definition of a factor:

$$P(x) = (x - c)q(x), \text{ where } q(x) \text{ is a polynomial}$$

Therefore, $P(c) = (c - c)q(c) = 0$, so c is a zero of the polynomial. Conversely, if c is a zero of the polynomial, then $P(c) = 0$. The Remainder Theorem states that the remainder when $P(x)$ is divided by $(x - c)$ is $P(c) = 0$. Therefore, $(x - c)$ is a factor of $P(x)$. The Factor Theorem can be used for both real and complex quadratic polynomials.

The quadratic formula:

$$x = \frac{-b \pm \sqrt{b^2 - 4ac}}{2a}$$

that solves $ax^2 + bx + c = 0$ can be used for both real and complex coefficients. It is proven by completing the square on the left-hand side of the equation $x^2 + \frac{b}{a}x = -\frac{c}{a}$, to obtain $\left(x + \frac{b}{2a}\right)^2 = \frac{b^2 - 4ac}{2a}$, which is then solved for x.

Solving Polynomial Inequalities

To solve a polynomial inequality, it must be in standard form, meaning it must be in descending order and the non-zero terms must all be on one side of the inequality. Then, the critical values are found by changing the inequality symbol to an equals sign and solving the equation. Next, a sign analysis chart is completed using the critical values. Such analysis involves splitting the number line into intervals using the critical values. A number is selected from each interval and plugged into the polynomial. The sign of the result will be the same as the sign for the entire interval. The intervals that satisfy the inequality are those that form the solution set. The answer can be written in interval notation. For example, the solution to $x^2 - 2x - 8 \geq 0$ is $\{x | x \leq 2 \text{ or } x \geq 4\}$ The critical values of $x^2 - 2x - 8 = 0$ are -2 and 4, which splits the number line into three intervals. The corresponding intervals that are greater than or equal to 0 when plugged into $x^2 - 2x - 8$ are those listed in the solution set.

Expanding Polynomials

Some polynomials may need to be expanded to identify the final solution—*polynomial expansion* means that parenthetical polynomials are multiplied out so that the parentheses no longer exist. The polynomials will be in the form $(a + b)^n$ where n is a whole number greater than 2. The expression can be simplified using the *distributive property*, which states that a number, variable, or polynomial that is multiplied by a polynomial in parentheses should be multiplied by each item in the parenthetical polynomial. Here's one example:

$$(a + b)^2 = (a + b)(a + b)$$

$$a^2 + ab + ab + b^2 = a^2 + 2ab + b^2$$

Here's another example to consider:

$$(a + b)^3 = (a + b)(a + b)(a + b)$$

$$(a^2 + ab + ab + b^2)(a + b)$$

$$(a^2 + 2ab + b^2)(a + b)$$

$$a^3 + 2a^2b + ab^2 + a^2b + 2ab^2 + b^3$$

$$a^3 + 3a^2b + 3ab^2 + b^3$$

Create Equations and Inequalities in One Variable

Linear expressions and equations are concise mathematical statements that can be written to model a variety of scenarios. Questions found pertaining to this topic will contain one variable only. A variable is an unknown quantity, usually denoted by a letter (*x, n, p,* etc.). In the case of linear expressions and

equations, the power of the variable (its exponent) is 1. A variable without a visible exponent is raised to the first power.

Writing Linear Expressions and Equations

A linear expression is a statement about an unknown quantity expressed in mathematical symbols. The statement "five times a number added to forty" can be expressed as $5x + 40$. A linear equation is a statement in which two expressions (at least one containing a variable) are equal to each other. The statement "five times a number added to forty is equal to ten" can be expressed as $5x + 40 = 10$. Real-world scenarios can also be expressed mathematically. Consider the following:

> Bob had $20 and Tom had $4. After selling 4 ice cream cones to Bob, Tom has as much money as Bob.

The cost of an ice cream cone is an unknown quantity and can be represented by a variable. The amount of money Bob has after his purchase is four times the cost of an ice cream cone subtracted from his original $20. The amount of money Tom has after his sale is four times the cost of an ice cream cone added to his original $4. This can be expressed as: $20 - 4x = 4x + 4$, where x represents the cost of an ice cream cone.

When expressing a verbal or written statement mathematically, it is key to understand words or phrases that can be represented with symbols. The following are examples:

Symbol	Phrase
$+$	added to, increased by, sum of, more than
$-$	decreased by, difference between, less than, take away
x	multiplied by, 3 (4, 5 . . .) times as large, product of
$\div$	divided by, quotient of, half (third, etc.) of
$=$	is, the same as, results in, as much as
$x, t, n, etc.$	a number, unknown quantity, value of

When graphing a linear function, note that the ratio of the change of the y coordinate to the change in the x coordinate is constant between any two points on the resulting line, no matter which two points are chosen. In other words, in a pair of points on a line, (x_1, y_1) and (x_2, y_2), with $x_1 \neq x_2$ so that the two points are distinct, then the ratio $\frac{y_2 - y_1}{x_2 - x_1}$ will be the same, regardless of which particular pair of points are chosen. This ratio, $\frac{y_2 - y_1}{x_2 - x_1}$, is called the *slope* of the line and is frequently denoted with the letter m. If slope m is positive, then the line goes upward when moving to the right, while if slope m is negative, then the line goes downward when moving to the right. If the slope is 0, then the line is called *horizontal*, and the y coordinate is constant along the entire line. In lines where the x coordinate is constant along the entire line, y is not actually a function of x. For such lines, the slope is not defined. These lines are called *vertical* lines.

Linear functions may take forms other than $y = ax + b$. The most common forms of linear equations are explained below:

- Standard Form: $Ax + By = C$, in which the slope is given by $m = \frac{-A}{B}$, and the y-intercept is given by $\frac{C}{B}$.

- Slope-Intercept Form: $y = mx + b$, where the slope is m and the y intercept is b.

- Point-Slope Form: $y - y_1 = m(x - x_1)$, where the slope is m and (x_1, y_1) is any point on the chosen line.

- Two-Point Form: $\frac{y-y_1}{x-x_1} = \frac{y_2-y_1}{x_2-x_1}$, where (x_1, y_1) and (x_2, y_2) are any two distinct points on the chosen line. Note that the slope is given by $m = \frac{y_2-y_1}{x_2-x_1}$.

- Intercept Form: $\frac{x}{x_1} + \frac{y}{y_1} = 1$, in which x_1 is the x-intercept and y_1 is the y-intercept.

These five ways to write linear equations are all useful in different circumstances. Depending on the given information, it may be easier to write one of the forms over another.

If $y = mx$, y is directly proportional to x. In this case, changing x by a factor changes y by that same factor. If $y = \frac{m}{x}$, y is inversely proportional to x. For example, if x is increased by a factor of 3, then y will be decreased by the same factor, 3.

Creating, Solving, or Interpreting Linear Inequalities in One Variable

Linear inequalities and linear equations are both comparisons of two algebraic expressions. However, unlike equations in which the expressions are equal to each other, linear inequalities compare expressions that are unequal. Linear equations typically have one value for the variable that makes the statement true. Linear inequalities generally have an infinite number of values that make the statement true. Exceptions to these last two statements are covered in Section 6.

Writing Linear Inequalities

Linear inequalities are a concise mathematical way to express the relationship between unequal values. More specifically, they describe in what way the values are unequal. A value could be greater than (>); less than (<); greater than or equal to (≥); or less than or equal to (≤) another value. The statement "five times a number added to forty is more than sixty-five" can be expressed as $5x + 40 > 65$. Common words and phrases that express inequalities are:

Symbol	Phrase
<	is under, is below, smaller than, beneath
>	is above, is over, bigger than, exceeds
≤	no more than, at most, maximum
≥	no less than, at least, minimum

Solving Linear Inequalities

When solving a linear inequality, the solution is the set of all numbers that makes the statement true. The inequality $x + 2 \geq 6$ has a solution set of 4 and every number greater than 4 (4.0001, 5, 12, 107,

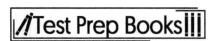

etc.). Adding 2 to 4 or any number greater than 4 would result in a value that is greater than or equal to 6. Therefore, $x \geq 4$ would be the solution set.

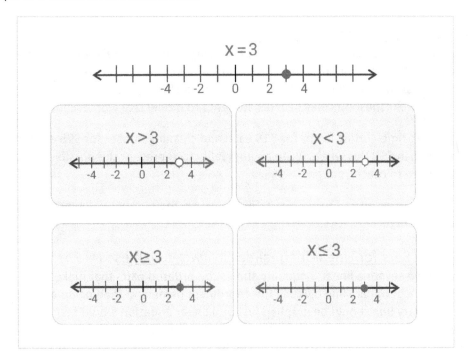

Solution sets for linear inequalities often will be displayed using a number line. If a value is included in the set ($\geq$ or $\leq$), there is a shaded dot placed on that value and an arrow extending in the direction of the solutions. For a variable $>$ or $\geq$ a number, the arrow would point right on the number line (the direction where the numbers increase); and if a variable is $<$ or $\leq$ a number, the arrow would point left (where the numbers decrease). If the value is not included in the set ($>$ or $<$), an open circle on that value would be used with an arrow in the appropriate direction.

Students may be asked to write a linear inequality given a graph of its solution set. To do so, they should identify whether the value is included (shaded dot or open circle) and the direction in which the arrow is pointing.

In order to algebraically solve a linear inequality, the same steps should be followed as in solving a linear equation. The inequality symbol stays the same for all operations EXCEPT when dividing by a negative number. If dividing by a negative number while solving an inequality, the relationship reverses (the sign flips). Dividing by a positive does not change the relationship, so the sign stays the same. In other words, $>$ switches to $<$ and vice versa. An example is shown below.

Solve $-2(x + 4) \leq 22$

Distribute: $-2x - 8 \leq 22$

Add 8 to both sides: $-2x \leq 30$

Divide both sides by -2: $x \geq 15$

Create Equations in Two or More Variables to Represent Relationships Between Quantities

Expressing Linear Inequalities in Two Variables

A linear inequality in two variables is a statement expressing an unequal relationship between those two variables. Typically written in slope-intercept form, the variable y can be greater than; less than; greater than or equal to; or less than or equal to a linear expression including the variable x. Examples include $y > 3x$ and $y \leq \frac{1}{2}x - 3$. Questions may instruct students to model real world scenarios such as:

> You work part-time cutting lawns for $15 each and cleaning houses for $25 each. Your goal is to make more than $90 this week. Write an inequality to represent the possible pairs of lawns and houses needed to reach your goal.

This scenario can be expressed as $15x + 25y > 90$ where x is the number of lawns cut and y is the number of houses cleaned.

Graphing Solution Sets for Linear Inequalities in Two Variables

A graph of the solution set for a linear inequality shows the ordered pairs that make the statement true. The graph consists of a boundary line dividing the coordinate plane and shading on one side of the boundary. The boundary line should be graphed just as a linear equation would be graphed. If the inequality symbol is $>$ or $<$, a dashed line can be used to indicate that the line is not part of the solution set. If the inequality symbol is $\geq$ or $\leq$, a solid line can be used to indicate that the boundary line is included in the solution set. An ordered pair (x, y) on either side of the line should be chosen to test in the inequality statement. If substituting the values for x and y results in a true statement ($15(3) + 25(2) > 90$), that ordered pair and all others on that side of the boundary line are part of the solution set. To indicate this, that region of the graph should be shaded. If substituting the ordered pair results in a false statement, the ordered pair and all others on that side are not part of the solution set.

Therefore, the other region of the graph contains the solutions and should be shaded.

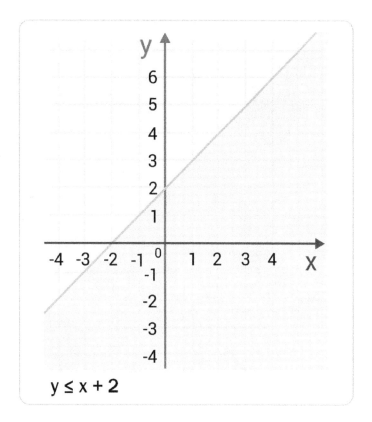

$$y \leq x + 2$$

A question may simply ask whether a given ordered pair is a solution to a given inequality. To determine this, the values should be substituted for the ordered pair into the inequality. If the result is a true statement, the ordered pair is a solution; if the result is a false statement, the ordered pair is not a solution.

Expressing Systems of Linear Inequalities in Two Variables

A system of linear inequalities consists of two linear inequalities making comparisons between two variables. Students may be given a scenario and asked to express it as a system of inequalities:

> A consumer study calls for at least 60 adult participants. It cannot use more than 25 men. Express these constraints as a system of inequalities.

This can be modeled by the system: $x + y \geq 60; x \leq 25$, where x represents the number of men and y represents the number of women. A solution to the system is an ordered pair that makes both inequalities true when substituting the values for x and y.

Graphing Solution Sets for Systems of Linear Inequalities in Two Variables

The solution set for a system of inequalities is the region of a graph consisting of ordered pairs that make both inequalities true. To graph the solution set, each linear inequality should first be graphed with appropriate shading. The region of the graph should be identified where the shading for the two inequalities overlaps. This region contains the solution set for the system.

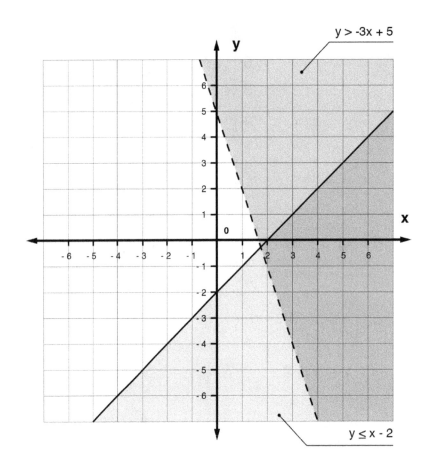

An ordered pair from the region of solutions can be selected to test in the system of inequalities.

Just as with manipulating linear inequalities in one variable, if dividing by a negative number in working with a linear inequality in two variables, the relationship reverses and the inequality sign should be flipped.

Expressing Systems of Two Linear Equations in Two Variables

A system of two linear equations in two variables is a set of equations that use the same variables, usually x and y. Here's a sample problem:

> An Internet provider charges an installation fee and a monthly charge. It advertises that two months of its offering costs $100 and six months costs $200. Find the monthly charge and the installation fee.

The two unknown quantities (variables) are the monthly charge and the installation fee. There are two different statements given relating the variables: two months added to the installation fee is $100; and six months added to the installation fee is $200. Using the variable x as the monthly charge and y as the installation fee, the statements can be written as the following: $2x + y = 100$; $6x + y = 200$. These two equations taken together form a system modeling the given scenario.

Solutions of a System of Two Linear Equations in Two Variables

A solution for a system of equations is an ordered pair that makes both equations true. One method for solving a system of equations is to graph both lines on a coordinate plane. If the lines intersect, the point of intersection is the solution to the system. Every point on a line represents an ordered pair that makes its equation true. The ordered pair represented by this point of intersection lies on both lines and therefore makes both equations true. This ordered pair should be checked by substituting its values into both of the original equations of the system. Note that given a system of equations and an ordered pair, the ordered pair can be determined to be a solution or not by checking it in both equations.

If, when graphed, the lines representing the equations of a system do not intersect, then the two lines are parallel to each other or they are the same exact line. Parallel lines extend in the same direction without ever meeting. A system consisting of parallel lines has no solution. If the equations for a system represent the same exact line, then every point on the line is a solution to the system. In this case, there would be an infinite number of solutions. A system consisting of intersecting lines is referred to as independent; a system consisting of parallel lines is referred to as inconsistent; and a system consisting of coinciding lines is referred to as dependent.

Here's some examples:

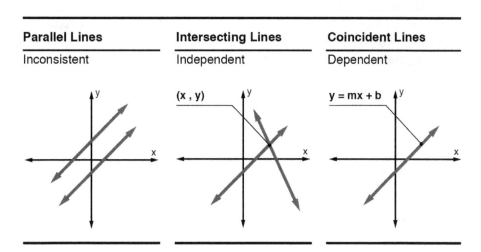

Represent Constraints by Equations or Inequalities and By Systems of Equations and/or Inequalities

Equations and Inequalities

The sum of a number and 5 is equal to -8 times the number. To find this unknown number, a simple equation can be written to represent the problem. Key words such as difference, equal, and times are used to form the following equation with one variable: $n + 5 = -8n$. When solving for n, opposite operations are used. First, n is subtracted from $-8n$ across the equals sign, resulting in $5 = -9n$. Then, -9 is divided on both sides, leaving $n = -\frac{5}{9}$. This solution can be graphed on the number line with a dot as shown below:

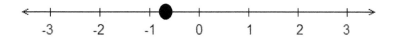

If the problem were changed to say, "The sum of a number and 5 is greater than -8 times the number," then an inequality would be used instead of an equation. Using key words again, *greater than* is represented by the symbol >. The inequality $n + 5 > -8n$ can be solved using the same techniques, resulting in $n < -\frac{5}{9}$. The only time solving an inequality differs from solving an equation is when a negative number is either multiplied times or divided by each side of the inequality. The sign must be switched in this case. For this example, the graph of the solution changes to the following graph because the solution represents all real numbers less than $-\frac{5}{9}$. Not included in this solution is $-\frac{5}{9}$ because it is a *less than* symbol, not *equal to*.

Equations and inequalities in two variables represent a relationship. Jim owns a car wash and charges $40 per car. The rent for the facility is $350 per month. An equation can be written to relate the number of cars Jim cleans to the money he makes per month. Let x represent the number of cars and y represent the profit Jim makes each month from the car wash. The equation $y = 40x - 350$ can be used to show Jim's profit or loss. Since this equation has two variables, the coordinate plane can be used to show the relationship and predict profit or loss for Jim. The following graph shows that Jim must wash at least nine cars to pay the rent, where $x = 9$. Anything nine cars and above yield a profit shown in the value on the y-axis.

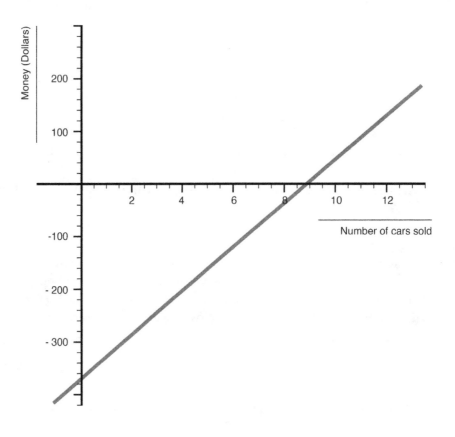

With a single equation in two variables, the solutions are limited only by the situation the equation represents. When two equations or inequalities are used, more constraints are added. For example, in a system of linear equations, there is often—although not always—only one answer. The point of intersection of two lines is the solution. For a system of inequalities, there are infinitely many answers.

The intersection of two solution sets gives the solution set of the system of inequalities. In the following graph, the darker shaded region is where two inequalities overlap. Any set of x and y found in that region satisfies both inequalities. The line with the positive slope is solid, meaning the values on that line are included in the solution. The line with the negative slope is dotted, so the coordinates on that line are not included.

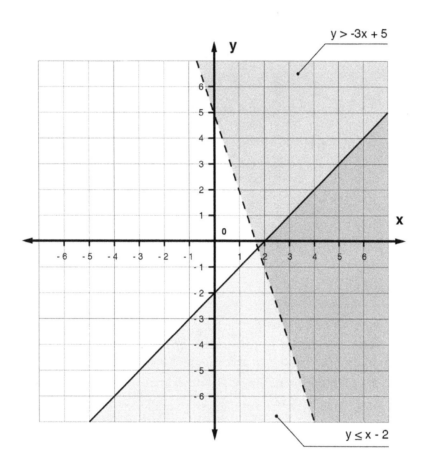

Formulas with two variables are equations used to represent a specific relationship. For example, the formula $d = rt$ represents the relationship between distance, rate, and time. If Bob travels at a rate of 35 miles per hour on his road trip from Westminster to Seneca, the formula $d = 35t$ can be used to represent his distance traveled in a specific length of time. Formulas can also be used to show different roles of the variables, transformed without any given numbers. Solving for r, the formula becomes $\frac{d}{t} = r$. The t is moved over by division so that *rate* is a function of distance and time.

Equivalent Forms of Functions

Equations in two variables can often be written in different forms to easily recognize a given trait of the function or its graph. Linear equations written in slope-intercept form allow for recognition of the slope and y-intercept; and linear equations written in standard form allow for identification of the x and y-

intercepts. Quadratic functions written in standard form allow for identification of the *y*-intercept and for easy calculation of outputs; and quadratic functions written in vertex form allow for identification of the function's minimum or maximum output and its graph's vertex. Polynomial functions written in factored form allow for identification of the zeros of the function.

The method of substituting the same inputs (*x*-values) into functions to determine if they produce the same outputs can reveal if functions are not equivalent (different outputs). However, corresponding inputs and outputs do not necessarily indicate equivalent functions.

It is possible for algebraic expressions and equations to be written that look completely different, yet are still equivalent. For instance, the expression $4(2x - 3) - 3x + 5$ is equivalent to the expression $5x - 7$. Given two algebraic expressions, it can be determined if they are equivalent by writing them in simplest form. Distribution should be used, if applicable, and like terms should be combined. Given two algebraic equations, it can be determined if they are equivalent by solving each for the same variable. Here are two sample equations to consider:

$$3x - 4y = 7 \text{ and } x + 2 = \frac{4}{3}y + 4\frac{1}{3}$$

To determine if they are equivalent, solving for x is required.

$3x - 4y = 7$ $x + 2 = \frac{4}{3}y + 4\frac{1}{3}$

$3x = 4y + 7$ $x = \frac{4}{3}y + 2\frac{1}{3}$

$x = \frac{4}{3}y + \frac{7}{3}$ $x = \frac{4}{3}y + 2\frac{1}{3}$

The equations are equivalent.

Converting To and From Radical Form

Algebraic expressions involving radicals ($\sqrt{\ }$, $\sqrt[3]{\ }$, etc.) can be written without the radical by using rational (fraction) exponents. For radical expressions, the value under the root symbol is called the radicand, and the type of root determines the index. For example, the expression $\sqrt{6x}$ has a radicand of 6x and index of 2 (it is a square root). If the exponent of the radicand is 1, then $\sqrt[n]{a} = a^{\frac{1}{n}}$ where n is the index. A number or variable without a power has an implied exponent of 1. For example, $\sqrt{6} = 6^{\frac{1}{2}}$ and $125^{\frac{1}{3}} = \sqrt[3]{125}$. For any exponent of the radicand:

$$\sqrt[n]{a^m} = \left(\sqrt[n]{a}\right)^m = a^{\frac{m}{n}}$$

For example:

$$64^{\frac{5}{3}} = \sqrt[3]{64^5} or \left(\sqrt[3]{64}\right)^5$$

$$(xy)^{\frac{2}{3}} = \sqrt[3]{(xy)^2} \text{ or } \left(\sqrt[3]{xy}\right)^2$$

Simplifying Expressions with Rational Exponents

When simplifying expressions with rational exponents, all basic properties for exponents hold true. When multiplying powers of the same base (same value with or without the same exponent), the exponents are added. For example:

$$x^{\frac{2}{7}} \times x^{\frac{3}{14}} = x^{\frac{1}{2}} \left(\frac{2}{7} + \frac{3}{14} = \frac{1}{2}\right)$$

When dividing powers of the same base, the exponents are subtracted. For example:

$$\frac{5^{\frac{2}{3}}}{5^{\frac{1}{2}}} = 5^{\frac{1}{6}} \left(\frac{2}{3} - \frac{1}{2} = \frac{1}{6}\right)$$

When raising a power to a power, the exponents are multiplied. For example:

$$\left(5^{\frac{1}{2}}\right)^4 = 5^2 \left(\frac{1}{2} \times 4 = 2\right)$$

When simplifying expressions with exponents, a number should never be raised to a power or a negative exponent. If a number has an integer exponent, its value should be determined. If the number has a rational exponent, it should be rewritten as a radical and the value determined if possible. A base with a negative exponent moves from the numerator to the denominator of a fraction (or vice versa) and is written with a positive exponent. For example, $x^{-3} = \frac{1}{x^3}$ and $\frac{2}{5x^{-2}} = \frac{2x^2}{5}$. The exponent of 5 is 1, and therefore the 5 does not move.

Here's a sample expression:

$$(27x^{-9})^{\frac{1}{3}}$$

After the implied exponents are noted, a power should be raised to a power by multiplying exponents, which yields:

$$27^{\frac{1}{3}} x^{-3}$$

Next, the negative exponent is eliminated by moving the base and power:

$$\frac{27^{\frac{1}{3}}}{x^3}$$

Then the value of the number is determined to a power by writing it in radical form:

$$\frac{\sqrt[3]{27}}{x^3}$$

Simplifying yields:

$$\frac{3}{x^3}$$

Explain Each Step in Solving a Simple Equation

Equations with one variable can be solved using the addition principle and multiplication principle. If $a = b$, then $a + c = b + c$, and $ac = bc$. Given the equation $2x - 3 = 5x + 7$, the first step is to combine the variable terms and the constant terms. Using the principles, expressions can be added and subtracted onto and off both sides of the equals sign, so the equation turns into $-10 = 3x$. Dividing by 3 on both sides through the multiplication principle with $c = \frac{1}{3}$ results in the final answer of:

$$x = \frac{-10}{3}$$

Some equations have a higher degree and are not solved by simply using opposite operations. When an equation has a degree of 2, completing the square is an option. For example, the quadratic equation $x^2 - 6x + 2 = 0$ can be rewritten by completing the square. The goal of completing the square is to get the equation into the form:

$$(x - p)^2 = q$$

Using the example, the constant term 2 first needs to be moved over to the opposite side by subtracting. Then, the square can be completed by adding 9 to both sides, which is the square of half of the coefficient of the middle term $-6x$. The current equation is:

$$x^2 - 6x + 9 = 7$$

The left side can be factored into a square of a binomial, resulting in:

$$(x - 3)^2 = 7$$

To solve for x, the square root of both sides should be taken, resulting in $(x - 3) = \pm\sqrt{7}$, and $x = 3 \pm \sqrt{7}$.

Other ways of solving quadratic equations include graphing, factoring, and using the quadratic formula. The equation $y = x^2 - 4x + 3$ can be graphed on the coordinate plane, and the solutions can be observed where it crosses the x-axis. The graph will be a parabola that opens up with two solutions at 1 and 3.

The equation can also be factored to find the solutions. The original equation, $y = x^2 - 4x + 3$ can be factored into:

$$y = (x - 1)(x - 3)$$

Setting this equal to zero, the x-values are found to be 1 and 3, just as on the graph. Solving by factoring and graphing are not always possible. The quadratic formula is a method of solving quadratic equations that always results in exact solutions. The formula is:

$$x = \frac{-b \pm \sqrt{b^2 - 4ac}}{2a}$$

A, b, and c are the coefficients in the original equation in standard form $y = ax^2 + bx + c$. For this example:

$$x = \frac{4 \pm \sqrt{(-4)^2 - 4(1)(3)}}{2(1)}$$

$$\frac{4 \pm \sqrt{16 - 12}}{2} = \frac{4 \pm 2}{2} = 1, 3$$

The expression underneath the radical is called the *discriminant*. Without working out the entire formula, the value of the discriminant can reveal the nature of the solutions. If the value of the discriminant $b^2 - 4ac$ is positive, then there will be two real solutions. If the value is zero, there will be one real solution. If the value is negative, the two solutions will be imaginary or complex. If the solutions are complex, it means that the parabola never touches the x-axis. An example of a complex solution can be found by solving the following quadratic:

$$y = x^2 - 4x + 8$$

By using the quadratic formula, the solutions are found to be:

$$x = \frac{4 \pm \sqrt{(-4)^2 - 4(1)(8)}}{2(1)}$$

$$\frac{4 \pm \sqrt{16 - 32}}{2} = \frac{4 \pm \sqrt{-16}}{2} = 2 \pm 2i$$

The solutions both have a real part, 2, and an imaginary part, $2i$.

Multistep One-Variable Linear Equations and Inequalities

Linear equations and linear inequalities are both comparisons of two algebraic expressions. However, unlike equations in which the expressions are equal, linear inequalities compare expressions that may be unequal. Linear equations typically have one value for the variable that makes the statement true. Linear inequalities generally have an infinite number of values that make the statement true.

When solving a linear equation, the desired result requires determining a numerical value for the unknown variable. If given a linear equation involving addition, subtraction, multiplication, or division, working backwards isolates the variable. Addition and subtraction are inverse operations, as are multiplication and division. Therefore, they can be used to cancel each other out.

The first steps to solving linear equations are distributing, if necessary, and combining any like terms on the same side of the equation. Sides of an equation are separated by an *equal* sign. Next, the equation is manipulated to show the variable on one side. Whatever is done to one side of the equation must be done to the other side of the equation to remain equal. Inverse operations are then used to isolate the variable and undo the order of operations backwards. Addition and subtraction are undone, then multiplication and division are undone.

For example, solve: $4(t-2) + 2t - 4 = 2(9 - 2t)$

Distributing: $4t - 8 + 2t - 4 = 18 - 4t$

Combining like terms: $6t - 12 = 18 - 4t$

Adding $4t$ to each side to move the variable: $10t - 12 = 18$

Adding 12 to each side to isolate the variable: $10t = 30$

Dividing each side by 10 to isolate the variable: $t = 3$

The answer can be checked by substituting the value for the variable into the original equation, ensuring that both sides calculate to be equal.

Linear inequalities express the relationship between unequal values. More specifically, they describe in what way the values are unequal. A value can be greater than (>), less than (<), greater than or equal to (≥), or less than or equal to (≤) another value. $5x + 40 > 65$ is read as *five times a number added to forty is greater than sixty-five.*

When solving a linear inequality, the solution is the set of all numbers that make the statement true. The inequality $x + 2 \geq 6$ has a solution set of 4 and every number greater than 4 (4.01; 5; 12; 107; etc.). Adding 2 to 4 or any number greater than 4 results in a value that is greater than or equal to 6. Therefore, $x \geq 4$ is the solution set.

To algebraically solve a linear inequality, follow the same steps as those for solving a linear equation. The inequality symbol stays the same for all operations *except* when multiplying or dividing by a negative number. If multiplying or dividing by a negative number while solving an inequality, the relationship reverses (the sign flips). In other words, > switches to < and vice versa. Multiplying or dividing by a positive number does not change the relationship, so the sign stays the same. An example is shown below.

Solve $-2x - 8 \leq 22$

Add 8 to both sides: $-2x \leq 30$

Divide both sides by -2: $x \geq -15$

Solutions of a linear equation or a linear inequality are the values of the variable that make a statement true. In the case of a linear equation, the solution set (list of all possible solutions) typically consists of a single numerical value. To find the solution, the equation is solved by isolating the variable. For example, solving the equation $3x - 7 = -13$ produces the solution $x = -2$. The only value for x which produces a true statement is -2. This can be checked by substituting -2 into the original equation to check that both sides are equal. In this case, $3(-2) - 7 = -13 \rightarrow -13 = -13$; therefore, -2 is a solution.

Although linear equations generally have one solution, this is not always the case. If there is no value for the variable that makes the statement true, there is no solution to the equation. Consider the equation:

$$x + 3 = x - 1$$

There is no value for *x* in which adding 3 to the value produces the same result as subtracting one from the value. Conversely, if any value for the variable makes a true statement, the equation has an infinite number of solutions. Consider the equation:

$$3x + 6 = 3(x + 2)$$

Any number substituted for *x* will result in a true statement (both sides of the equation are equal).

By manipulating equations like the two above, the variable of the equation will cancel out completely. If the remaining constants express a true statement (ex. $6 = 6$), then all real numbers are solutions to the equation. If the constants left express a false statement (ex. $3 = -1$), then no solution exists for the equation.

Solving a linear inequality requires all values that make the statement true to be determined. For example, solving $3x - 7 \geq -13$ produces the solution $x \geq -2$. This means that -2 and any number greater than -2 produces a true statement. Solution sets for linear inequalities will often be displayed using a number line. If a value is included in the set ($\geq$ or $\leq$), a shaded dot is placed on that value and an arrow extending in the direction of the solutions. For a variable > or $\geq$ a number, the arrow will point right on a number line, the direction where the numbers increase. If a variable is < or $\leq$ a number, the arrow will point left on a number line, which is the direction where the numbers decrease. If the value is not included in the set (> or <), an open (unshaded) circle on that value is used with an arrow in the appropriate direction.

Like this:

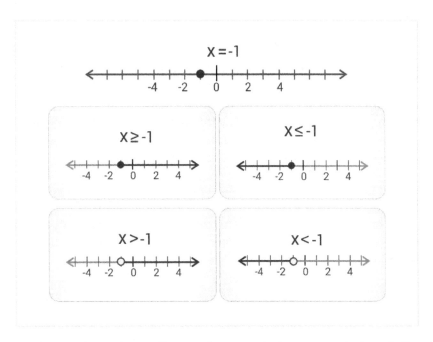

Similar to linear equations, a linear inequality may have a solution set consisting of all real numbers, or can contain no solution. When solved algebraically, a linear inequality in which the variable cancels out and results in a true statement (ex. $7 \geq 2$) has a solution set of all real numbers. A linear inequality in which the variable cancels out and results in a false statement (ex. $7 \leq 2$) has no solution.

Solve Linear Equations and Inequalities in One Variable

An *equation* is an expression that uses an equals sign to demonstrate that two quantities are equal to one another, such as the expression:

$$x^2 - x = -4x + 3$$

Solving an equation means to find all possible values that *x* can take which make the equation true.

Given an equation where one side is a polynomial in the variable *x* and the other side is zero, the solutions are also called the *roots* or *zeros* of the equation.

To solve an equation, the equation needs to be modified to determine the solution. Starting with an equation $a = b$, the following are also true equations:

$$a + c = b + c$$

$$a - c = b - c$$

$$ac = bc$$

$$a/c = b/c \text{ (provided that } c \text{ is not 0)}$$

$$a^c = b^c$$

$$\sqrt{a} = \pm\sqrt{b}$$

The following rule is important to remember when solving equations:

$$\text{If } ab = 0, \text{ then } a = 0 \text{ or } b = 0.$$

Sometimes, instead of an equation, an *inequality* is used to indicate that one quantity is less than another (or greater than another). They may specify that the two quantities might also be equal to each other. If the quantities are not allowed to equal one another, the expression is a *strict inequality*. For example, $x + 3 \leq 5$ is an inequality, while $7 - 2x > 1$ is a strict inequality.

A *solution set* is a collection of all values of a variable that solve an equation or an inequality. For inequalities, this can be illustrated on a number line by shading in the part of the number line that satisfies the inequality. An open circle on the number line indicates that one gets arbitrarily close to that point, but cannot actually touch that point while remaining in the solution set. For example, to graph the solution set for the inequality $x > 2$, it would look like the following:

Solution Set x>2

Solving equations in one variable is the process of For example, in $3x - 7 = 20$, the variable x needs to be isolated. Using opposite operations, the -7 is moved to the right side of the equation by adding seven to both sides: $3x - 7 + 7 = 20 + 7$, resulting in $3x = 27$. Dividing by three on each side, $\frac{3x}{3} = \frac{27}{3}$, results in isolation of the variable. It is important to note that if an operation is performed on one side of the equals sign, it has to be performed on the other side to maintain equality. The solution is found to be $x = 9$. This solution can be checked for accuracy by plugging $x=7$ in the original equation. After simplifying the equation, $20 = 20$ is found, which is a true statement.

When solving radical and rational equations, extraneous solutions must be accounted for when finding the answers. For example, the equation $\frac{x}{x-5} = \frac{3x}{x+3}$ has two values that create a 0 denominator: $x \neq 5, -3$. When solving for x, these values must be considered because they cannot be solutions. In the given equation, solving for x can be done using cross-multiplication, yielding the equation:

$$x(x + 3) = 3x(x - 5)$$

Distributing results in the quadratic equation yields $x^2 + 3x = 3x^2 - 15x$; therefore, all terms must be moved to one side of the equals sign. This results in $2x^2 - 18x = 0$, which in factored form is $2x(x - 9) = 0$. Setting each factor equal to zero, the apparent solutions are $x = 0$ and $x = 9$. These two solutions are neither 5 nor -3, so they are viable solutions. Neither 0 nor 9 create a 0 denominator in the original equation.

A similar process exists when solving radical equations. One must check to make sure the solutions are defined in the original equations. Solving an equation containing a square root involves isolating the root and then squaring both sides of the equals sign. Solving a cube root equation involves isolating the radical and then cubing both sides. In either case, the variable can then be solved for because there are no longer radicals in the equation.

Solving Equations in One Variable
Solving equations in one variable is the process of isolating a variable on one side of the equation. The letters in an equation and any numbers attached to them are the variables as they stand for unknown quantities that you are trying to solve for. *X* is commonly used as a variable, though any letter can be used. For example, in $3x - 7 = 20$, the variable is $3x$, and it needs to be isolated. The numbers (also called constants) are -7 and 20. That means $3x$ needs to be on one side of the equals sign (either side is fine), and all the numbers need to be on the other side of the equals sign.

To accomplish this, the equation must be manipulated by performing opposite operations of what already exists. Remember that addition and subtraction are opposites and that multiplication and division are opposites. Any action taken to one side of the equation must be taken on the other side to maintain equality.

So, since the 7 is being subtracted, it can be moved to the right side of the equation by adding seven to both sides:

$$3x - 7 = 20$$
$$3x - 7 + 7 = 20 + 7$$
$$3x = 27$$

Now that the variable $3x$ is on one side and the constants (now combined into one constant) are on the other side, the 3 needs to be moved to the right side. 3 and x are being multiplied together, so 3 then needs to be divided from each side.

$$\frac{3x}{3} = \frac{27}{3}$$

$$x = 9$$

Now that x has been completely isolated, we know its value.

The solution is found to be $x = 9$. This solution can be checked for accuracy by plugging $x = 9$ in the original equation. After simplifying the equation, $20 = 20$ is found, which is a true statement:

$$3 \times 9 - 7 = 20$$

$$27 - 7 = 20$$

$$20 = 20$$

Equations that require solving for a variable (*algebraic equations*) come in many forms. Here are some more examples:

No number attached to the variable:

$$x + 8 = 20$$

$$x + 8 - 8 = 20 - 8$$

$$x = 12$$

Fraction in the variable:

$$\frac{1}{2}z + 24 = 36$$

$$\frac{1}{2}z + 24 - 24 = 36 - 24$$

$$\frac{1}{2}z = 12$$

Now we multiply the fraction by its inverse:

$$\frac{2}{1} \times \frac{1}{2}z = 12 \times \frac{2}{1}$$

$$z = 24$$

Multiple instances of x:

$$14x + x - 4 = 3x + 2$$

All instances of x can be combined.

$$15x - 4 = 3x + 2$$

$$15x - 4 + 4 = 3x + 2 + 4$$

$$15x = 3x + 6$$

$$15x - 3x = 3x + 6 - 3x$$

$$12x = 6$$

$$\frac{12x}{12} = \frac{6}{12}$$

$$x = \frac{1}{2}$$

Solving an Equation in One Variable that Contains Radicals or Contains the Variable in the Denominator of a Fraction

Equations with radicals containing numbers only as the radicand are solved the same way that an equation without a radical would be. For example, $3x + \sqrt{81} = 45$ would be solved using the same steps as if solving:

$$2x + 4 = 12$$

Radical equations are those in which the variable is part of the radicand. For example, $\sqrt{5x+1} - 6 = 0$ and $\sqrt{x-3} + 5 = x$ would be considered radical equations.

Radical Equations

To solve a radical equation, the radical should be isolated and both sides of the equation should be raised to the same power to cancel the radical. Raising both sides to the second power will cancel a square root, raising to the third power will cancel a cube root, etc. To solve $\sqrt{5x+1} - 6 = 0$, the radical should be isolated first:

$$\sqrt{5x+1} = 6$$

Then both sides should be raised to the second power:

$$(\sqrt{5x+1})^2 = (6)^2 \rightarrow 5x + 1 = 36$$

Lastly, the linear equation should be solved:

$$x = 7$$

Radical Equations with Extraneous Solutions

If a radical equation contains a variable in the radicand and a variable outside of the radicand, it must be checked for extraneous solutions. An extraneous solution is one obtained by following the proper process for solving an equation but does not "check out" when substituted into the original equation. Here's a sample equation:

$$\sqrt{x-3} + 5 = x$$

Isolating the radical yields:

$$\sqrt{x-3} = x - 5$$

Next, both sides should be squared to cancel the radical:

$$\left(\sqrt{x-3}\right)^2 = (x-5)^2$$

$$x - 3 = (x-5)(x-5)$$

The binomials should be multiplied:

$$x - 3 = x^2 - 10x + 25$$

The quadratic equation is then solved:

$$0 = x^2 - 11x + 28$$

$$0 = (x-7)(x-4)$$

$$x - 7 = 0; x - 4 = 0$$

$$x = 7 \text{ or } x = 4$$

To check for extraneous solutions, each answer can be substituted, one at a time, into the original equation. Substituting 7 for *x*, results in $7 = 7$. Therefore, 7 is a solution. Substituting 4 for *x* results in $6 = 4$. This is false; therefore, 4 is an extraneous solution.

Equations with a Variable in the Denominator of a Fraction

For equations with variables in the denominator, if the equation contains two rational expressions (on opposite sides of the equation, or on the same side and equal to zero), it can be solved like a proportion. Here's an equation to consider:

$$\frac{5}{2x-2} = \frac{15}{x^2-1}$$

First, cross-multiplying yields:

$$5(x^2-1) = 15(2x-2)$$

Distributing yields:

$$5x^2 - 5 = 30x - 30$$

In solving the quadratic equation, it is determined that $x = 1$ or $x = 5$. Solutions must be checked to see if they are extraneous. Extraneous solutions either produce a false statement when substituted into the original equation or create a rational expression with a denominator of zero (dividing by zero is

undefined). Substituting 5 into the original equation produces $\frac{5}{8} = \frac{5}{8}$; therefore, 5 is a solution. Substituting 1 into the original equation results in both denominators equal to zero; therefore, 1 is an extraneous solution.

If an equation contains three or more rational expressions: the least common denominator (LCD) needs to be found for all the expressions, then both sides of the equation should be multiplied by the LCD. The LCD consists of the lowest number that all coefficients divide evenly into and for every variable, the highest power of that variable. Here's a sample equation:

$$\frac{3}{5x} - \frac{4}{3x} = \frac{1}{3}$$

The LCD would be 15x. Both sides of the equation should be multiplied by 15x:

$$15x \left(\frac{3}{5x} - \frac{4}{3x} \right)$$

$$15x \left(\frac{1}{3} \right)$$

$$\frac{45x}{5x} - \frac{60x}{3x} = \frac{15x}{3}$$

$$9 - 20 = 5x$$

$$x = -2\frac{1}{2}$$

Any extraneous solutions should be identified.

Solve Quadratic Equations in One Variable

Quadratic Models
A quadratic function can be written in the standard form:

$$y = ax^2 + bx + c$$

It can be represented by a u-shaped graph called a parabola. For a quadratic function where the value of *a* is positive, as the inputs increase, the outputs increase until a certain value (maximum of the function) is reached. As inputs increase past the value that corresponds with the maximum output, the relationship reverses and the outputs decrease. For a quadratic function where *a* is negative, as the inputs increase, the outputs (1) decrease, (2) reach a maximum, and (3) then increase.

Consider a ball thrown straight up into the air. As time passes, the height of the ball increases until it reaches its maximum height. After reaching the maximum height, as time increases, the height of the ball decreases (it is falling toward the ground). This relationship can be expressed as a quadratic function where time is the input (*x*) and the height of the ball is the output (*y*).

Given a scenario that can be modeled by a quadratic function, to write its equation, the following is needed: its vertex and any other ordered pair; or any three ordered pairs for the function. Given three ordered pairs, they should be substituted into the general form ($y = ax^2 + bx + c$) to create a system of three equations. For example, given the ordered pairs (2, 3), (3, 13), and (4, 29), it yields:

$$3 = a(2)2 + b(2) + c \rightarrow 4a + 2b + c = 3$$

$$13 = a(3)2 + b(3) + c \rightarrow 9a + 3b + c = 13$$

$$29 = a(4)2 + b(4) + c \rightarrow 16a + 24b + c = 29$$

The values for a, b, and c in the system can be found and substituted into the general form to write the equation of the function. In this case, the equation is:

$$y = 3x^2 - 5x + 1$$

Exponential Models

Exponential functions can be written in the form:

$$y = a \times b^x$$

Scenarios involving growth and decay can be modeled by exponential functions.

The equation for an exponential function can be written given the y-intercept (a) and the growth rate (b). The y-intercept is the output (y) when the input (x) equals zero. It can be thought of as an "original value" or starting point. The value of b is the rate at which the original value increases ($b > 1$) or decreases ($b < 1$). Suppose someone deposits $1200 into a bank account that accrues 1% interest per month. The y-intercept, a, would be $1200, while the growth rate, b, would be 1.01 (100% of the original value + 1% interest). This scenario could be written as the exponential function $y = 1200 \times 1.01^x$, where x represents the number of months since the deposit and y represents money in the account.

Given a scenario that models an exponential function, the equation can also be written when provided two ordered pairs.

Solving a Quadratic Equation

A *quadratic equation* is an equation in the form:

$$ax^2 + bx + c = 0$$

There are several methods to solve such equations. The easiest method will depend on the quadratic equation in question.

A quadratic equation can have two, one, or zero real solutions. Depending on its structure, a quadratic equation can be solved by (1) factoring, (2) taking square roots, or (3) using the quadratic formula.

Solving Quadratic Equations by Factoring

It sometimes is possible to solve quadratic equations by manually *factoring* them. This means rewriting them in the form:

$$(x + A)(x + B) = 0$$

If this is done, then they can be solved by remembering that when $ab = 0$, either a or b must be equal to zero. Therefore, to have $(x + A)(x + B) = 0$, $(x + A) = 0$ or $(x + B) = 0$ is needed. These equations have the solutions $x = -A$ and $x = -B$, respectively.

In order to factor a quadratic equation, note that:

$$(x + A)(x + B) = x^2 + (A + B)x + AB$$

So, if an equation is in the form $x^2 + bx + c$, two numbers, A and B, need to be found that will add up to give us b, and multiply together to give us c.

As an example, consider solving the equation:

$$-3x^2 + 6x + 9 = 0$$

Start by dividing both sides by -3, leaving:

$$x^2 - 2x - 3 = 0$$

Now, notice that $1 - 3 = -2$, and also that:

$$(1)(-3) = -3$$

This means the equation can be factored into:

$$(x + 1)(x - 3) = 0$$

Now, solve $(x + 1) = 0$ and $(x - 3) = 0$ to get $x = -1$ and $x = 3$ as the solutions.

It is useful when trying to factor to remember that:

$$x^2 + 2xy + y^2 = (x + y)^2$$
$$x^2 - 2xy + y^2 = (x - y)^2$$

And

$$x^2 - y^2 = (x + y)(x - y)$$

To recap, to solve a quadratic equation by factoring, the equation should first be manipulated to set the quadratic expression equal to zero. Next, the quadratic expression should be factored using the appropriate method(s). Then each factor should be set equal to zero. If two factors multiply to equal zero, then one or both factors must equal zero. Finally, each equation should be solved. Here's a sample:

$$x^2 - 10 = 3x - 6$$

The expression should be set equal to zero: $x^2 - 3x - 4 = 0$. The expression should be factored:

$$(x - 4)(x + 1) = 0$$

Each factor should be set equal to zero: $x - 4 = 0$; $x + 1 = 0$. Solving yields $x = 4$ or $x = -1$.

Solving Quadratic Equations by Taking Square Roots
If a quadratic equation does not have a linear term (variable to the first power), it can be solved by taking square roots. This means x^2 needs to be isolated and then the square root of both sides of the equation should be isolated. There will be two solutions because square roots can be positive or negative. ($\sqrt{4}$ = 2 or -2 because $2 \times 2 = 4$ and $-2 \times -2 = 4$.) Here's a sample equation:

$$3x^2 - 12 = 0$$

Isolating x^2 yields $x^2 = 4$. The square root of both sides is then solved: $x = 2$ or -2.

The Quadratic Formula
If there are no obvious ways to factor the quadratic equation, solutions can still be found by using the *quadratic formula*.

The quadratic formula is:

$$x = \frac{-b \pm \sqrt{b^2 - 4ac}}{2a}$$

This method will always work, although it sometimes can take longer than factoring by hand, if the factors are easy to guess. Using the standard form $ax^2 + bx + c = 0$, plug the values of *a*, *b*, and *c* from the equation into the formula and solve for x. There will either be two answers, one answer, or no real answer. No real answer comes when the value of the discriminant, the number under the square root, is a negative number. Since there are no real numbers that square to get a negative, the answer will be no real roots.

Here is an example of solving a quadratic equation using the quadratic formula. Suppose the equation to solve is:

$$-2x^2 + 3x + 1 = 0$$

There is no obvious way to factor this, so the quadratic formula is used, with $a = -2, b = 3, c = 1$. After substituting these values into the quadratic formula, it yields this:

$$x = \frac{-3 \pm \sqrt{3^2 - 4(-2)(1)}}{2(-2)}$$

This can be simplified to obtain:

$$\frac{3 \pm \sqrt{9 + 8}}{4}$$

or

$$\frac{3 \pm \sqrt{17}}{4}$$

Consider another example. Again, remember that to solve using the quadratic equation, the equation must be in the form:

$$ax^2 + bx + c = 0$$

Again, the quadratic formula is:

$$x = \frac{-b \pm \sqrt{b^2 - 4ac}}{2a}$$

(The $\pm$ symbol indicates that two calculations are necessary, one using $+$ and one using $-$.) Here's a sample equation:

$$3x^2 - 2x = 3x + 2$$

First, the quadratic expression should be set equal to zero:

$$3x^2 - 5x - 2 = 0$$

Then the values are substituted for $a(3)$, $b(-5)$, and $c(-2)$ into the formula:

$$x = \frac{-(-5) \pm \sqrt{(-5)^2 - 4(3)(-2)}}{2(3)}$$

Simplification yields:

$$x = \frac{5 \pm \sqrt{49}}{6} \rightarrow x = \frac{5 \pm 7}{6}$$

Calculating two values for x using $+$ and $-$ yields:

$$x = \frac{5 + 7}{6}; x = \frac{5 - 7}{6}$$

Simplification yields:

$$x = 2 \text{ or } -\frac{1}{3}.$$

Just as with any equation, solutions should be checked by substituting the value into the original equation.

Challenges can be encountered when asked to find a quadratic equation with specific roots. Given roots A and B, a quadratic function can be constructed with those roots by taking:

$$(x - A)(x - B)$$

So, in constructing a quadratic equation with roots $x = -2, 3$, it would result in:

$$(x + 2)(x - 3) = x^2 - x - 6$$

Multiplying this by a constant also could be done without changing the roots.

Solve Systems of Linear Equations Exactly and Approximately

Sometimes, a problem involves multiple variables and multiple equations that must all hold true at the same time. There are a few basic rules to keep in mind when solving systems of equations.

A single equation can be changed by doing the same operation to both sides, just as with one equation.

If one of the equations gives an expression for one of the variables in terms of other variables and constants, the expression can be substituted into the other equation, in place of the variable. This means the other equations will have one less variable in them.

If two equations are in the form of $a = b$ and $c = d$, then a new equation can be formed by adding the equations together, $a + c = b + d$, or subtracting the equations:

$$a - c = b - d$$

This can eliminate one of the variables from an equation.

The general approach is to find a way to change one of the equations so that one variable is isolated, and then substitute that value (or expression) for the variable into the other equations.

The simplest case is a *linear system of two equations*, which has the form $ax + by = c, dx + ey = f$.

To solve linear systems of equations, use the same process to solve one equation in order to isolate one of the variables. Here's an example, using the linear system of equations:

$$2x - 3y = 2, 4x + 4y = 3$$

The first equation is multiplied on both sides by -2, which gives:

$$-4x + 6y = -4$$

Adding this equation to the second equation will allow cancellation of the *x* term:

$$4x + 4y - 4x + 6y = 3 - 4$$

The result can be simplified to get $10y = -1$, which simplifies to:

$$y = -\frac{1}{10}$$

The solution can be substituted into either of the original equations to find a value for *x*. Using the first equation:

$$2x - 3\left(-\frac{1}{10}\right) = 2$$

This simplifies to $2x + \frac{3}{10} = 2$, then to $2x = \frac{17}{10}$, and finally $x = \frac{17}{20}$.

The final solution is $x = \frac{17}{20}, y = -\frac{1}{10}$.

To check the validity of the answer, both solutions can be substituted into either original equation, which should result in a true statement.

An alternative way to solve this system would be to solve the first equation to get an expression for *y* in terms of *x*.

Subtracting $2x$ from both sides results in $-3y = 2 - 2x$.

Dividing both sides by -3 would be $y = \frac{2}{3}x - \frac{2}{3}$.

Then, this expression can be substituted into the second equation, getting:

$$4x + 4\left(\frac{2}{3}x - \frac{2}{3}\right) = 3$$

This only involves the variable *x*, which can now be solved. Once the value for *x* is obtained, it can be substituted into either equation to solve for *y*.

There is one important issue to note here. If one of the equations in the system can be made to look identical to another equation, then it is *redundant*. The set of solutions will then be all pairs that satisfy the other equation.

For instance, in the system of equations, $2x - y = 1, -4x + 2y = -2$, the second equation can be made into the first equation by dividing both sides by -2. Thus, the solution set will be all pairs satisfying $2x - y = 1$, which simplifies to:

$$y = 2x - 1$$

For a pair of linear equations, the simplest way to see if one equation is redundant is to rewrite each equation to the form:

$$ax + by = c$$

If one equation can be obtained from the other in this form by multiplying both sides by some constant, then the equations are redundant, and the answer to the system would be all real numbers.

It is also possible for the two equations to be *inconsistent*, which occurs when the two equations can be made into the form $ax + by = c, ax + by = d$, with *c* and *d* being different numbers. The two equations are inconsistent if, while trying to solve them, it is determined that an equation is false, such as $3 = 2$. This result shows that there are no solutions to that system of equations.

For linear systems of two equations with two variables, there will always be a single solution unless one of the two equations is redundant or the equations are inconsistent, in which case there are no solutions.

Solving a System of One Linear Equation and One Quadratic Equation

A system of equations consists of two variables in two equations. A solution to the system is an ordered pair (x, y) that makes both equations true. When displayed graphically, a solution to a system is a point of intersection between the graphs of the equations. When a system consists of one linear equation and one quadratic equation, there may be one, two, or no solutions.

If the line and parabola intersect at two points, there are two solutions to the system; if they intersect at one point, there is one solution; if they do not intersect, there is no solution.

Systems with One Linear Equation and One Quadratic Equation

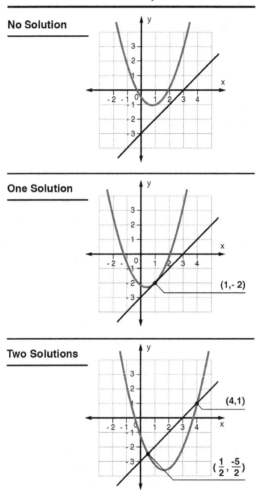

One method for solving a system of one linear equation and one quadratic equation is to graph both functions and identify point(s) of intersection. This, however, is not always practical. Graph paper may not be available or the intersection points may not be easily identified. Solving the system algebraically involves using the substitution method. Consider the following system:

$$y = x^2 + 9x + 11$$

$$y = 2x - 1$$

The equivalent value of y should be substituted from the linear equation ($2x - 1$) into the quadratic equation. The resulting equation is:

$$2x - 1 = x^2 + 9x + 11$$

Next, this quadratic equation should be solved using the appropriate method: factoring, taking square roots, or using the quadratic formula. Solving this quadratic equation by factoring results in $x = -4$ or $x = -3$. Next, the corresponding y-values should be found by substituting the x-values into the original linear equation:

$$y = 2(-4) - 1$$

$$y = 2(-3) - 1$$

The solutions should be written as ordered pairs: (-4, -9) and (-3, -7). Finally, the possible solutions should be checked by substituting each into both of the original equations. In this case, both solutions "check out."

Understand that the Graph of an Equation in Two Variables is the Set of All its Solutions Plotted in the Coordinate Plane

A graph can shift in many ways. To shift it horizontally, a constant can be added to all the x variables. Replacing x with $(x + a)$ will shift the graph to the left by a. If a is negative, this shifts the graph to the right. Similarly, vertical shifts occur by adding a constant to each of the y variables. Replacing y by $(y + a)$ will shift the graph up by a. If a is negative, then it shifts the graph down.

A graph can also stretch and shrink the graph in the horizontal and vertical directions. To stretch by a (positive) factor of k horizontally, all instances of x are replaced with $\frac{x}{k}$. To stretch vertically by k, all instances of y are replaced with $\frac{y}{k}$.

The graph can be reflected over the y-axis by replacing all instances of x with $(-x)$. The graph can also be reflected over the x-axis by replacing all instances of y with $(-y)$.

Graphing Quadratic Functions

The standard form of a quadratic function is:

$$y = ax^2 + bx + c$$

The graph of a quadratic function is a u-shaped (or upside down u) curve, called a parabola, which is symmetric about a vertical line (axis of symmetry). To graph a parabola, its vertex (high or low point for the curve) and at least two points on each side of the axis of symmetry need to be determined.

Given a quadratic function in standard form, the axis of symmetry for its graph is the line:

$$x = -\frac{b}{2a}$$

The vertex for the parabola has an x-coordinate of $-\frac{b}{2a}$. To find the y-coordinate for the vertex, the calculated x-coordinate needs to be substituted. To complete the graph, two different x-values need to be selected and substituted into the quadratic function to obtain the corresponding y-values. This will give two points on the parabola. These two points and the axis of symmetry are used to determine the two points corresponding to these. The corresponding points are the same distance from the axis of symmetry (on the other side) and contain the same *y*-coordinate. Plotting the vertex and four other points on the parabola allows for constructing the curve.

Quadratic Function

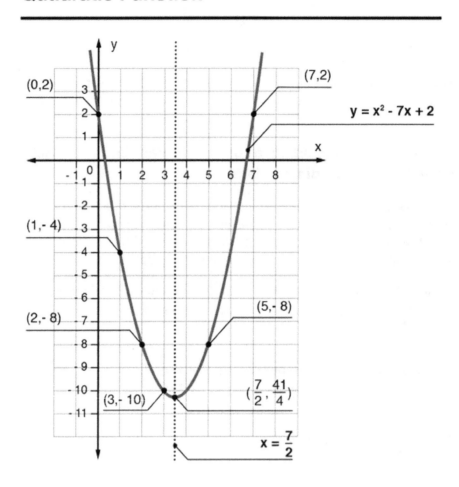

Graphing Exponential Functions

Exponential functions have a general form of $y = a \times b^x$. The graph of an exponential function is a curve that slopes upward or downward from left to right. The graph approaches a line, called an asymptote, as *x* or *y* increases or decreases. To graph the curve for an exponential function, *x*-values are selected and then substituted into the function to obtain the corresponding *y*-values. A general rule of thumb is to select three negative values, zero, and three positive values. Plotting the seven points on the graph for an exponential function should allow for constructing a smooth curve through them.

Functions

Understand that a Function from One Set to Another Set Assigns to Each Element of the Domain Exactly One Element of the Range

First, it's important to understand the definition of a *relation*. Given two variables, x and y, which stand for unknown numbers, a *relation* between x and y is an object that splits all of the pairs (x, y) into those for which the relation is true and those for which it is false. For example, consider the relation of $x^2 = y^2$. This relationship is true for the pair (1, 1) and for the pair (-2, 2), but false for (2, 3). Another example of a relation is $x \leq y$. This is true whenever x is less than or equal to y.

A *function* is a special kind of relation where, for each value of x, there is only a single value of y that satisfies the relation. So, $x^2 = y^2$ is *not* a function because in this case, if x is 1, y can be either 1 or -1: the pair (1, 1) and (1, -1) both satisfy the relation. More generally, for this relation, any pair of the form $(a, \pm a)$ will satisfy it. On the other hand, consider the following relation:

$$y = x^2 + 1$$

This is a function because for each value of x, there is a unique value of y that satisfies the relation. Notice, however, there are multiple values of x that give us the same value of y. This is perfectly acceptable for a function. Therefore, y is a function of x.

To determine if a relation is a function, check to see if every x value has a unique corresponding y value.

A function can be viewed as an object that has x as its input and outputs a unique y-value. It is sometimes convenient to express this using *function notation*, where the function itself is given a name, often f. To emphasize that f takes x as its input, the function is written as $f(x)$. In the above example, the equation could be rewritten as:

$$f(x) = x^2 + 1$$

To write the value that a function yields for some specific value of x, that value is put in place of x in the function notation. For example, $f(3)$ means the value that the function outputs when the input value is 3. If $f(x) = x^2 + 1$, then:

$$f(3) = 3^2 + 1 = 10$$

A function can also be viewed as a table of pairs (x, y), which lists the value for y for each possible value of x.

The set of all possible values for x in $f(x)$ is called the *domain* of the function, and the set of all possible outputs is called the *range* of the function. Note that usually the domain is assumed to be all real numbers, except those for which the expression for $f(x)$ is not defined, unless the problem specifies otherwise. An example of how a function might not be defined is in the case of $f(x) = \frac{1}{x+1}$, which is not defined when $x = -1$ (which would require dividing by zero). Therefore, in this case the domain would be all real numbers except $x = -1$.

If y is a function of x, then x is the *independent variable* and y is the *dependent variable*. This is because in many cases, the problem will start with some value of x and then see how y changes depending on this starting value.

Domain and Range

The domain and range of a function can be found visually by its plot on the coordinate plane. In the function $f(x) = x^2 - 3$, for example, the domain is all real numbers because the parabola stretches as far left and as far right as it can go, with no restrictions. This means that any input value from the real number system will yield an answer in the real number system. For the range, the inequality $y \geq -3$ would be used to describe the possible output values because the parabola has a minimum at $y = -3$. This means there will not be any real output values less than -3 because -3 is the lowest value it reaches on the y-axis.

These same answers for domain and range can be found by observing a table. The table below shows that from input values $x = -1$ to $x = 1$, the output results in a minimum of -3. On each side of $x = 0$, the numbers increase, showing that the range is all real numbers greater than or equal to -3.

x (domain/input)	y (range/output)
-2	1
-1	-2
0	-3
-1	-2
2	1

Use Function Notation, Evaluate Functions for Inputs in Their Domains, and Interpret Statements that Use Function Notation

A function is called *algebraic* if it is built up from polynomials by adding, subtracting, multiplying, dividing, and taking radicals. This means that, for example, the variable can never appear in an exponent. Thus, polynomials and rational functions are algebraic, but exponential functions are not algebraic. It turns out that logarithms and trigonometric functions are not algebraic either.

A function of the form $f(x) = a_n x^n + a_{n-1} x^{n-1} + a_{n-2} x^{n-2} + \cdots + a_1 x + a_0$ is called a *polynomial function*. The value of n is called the *degree* of the polynomial. In the case where $n = 1$, it is called a *linear function*. In the case where $n = 2$, it is called a *quadratic function*. In the case where $n = 3$, it is called a *cubic function*.

When n is even, the polynomial is called *even*, and not all real numbers will be in its range. When n is odd, the polynomial is called *odd*, and the range includes all real numbers.

The graph of a quadratic function $f(x) = ax^2 + bx + c$ will be a parabola. To see whether or not the parabola opens up or down, it's necessary to check the coefficient of x^2, which is the value a.

If the coefficient is positive, then the parabola opens upward. If the coefficient is negative, then the parabola opens downward.

The quantity $D = b^2 - 4ac$ is called the *discriminant* of the parabola. If the discriminant is positive, then the parabola has two real zeros. If the discriminant is negative, then it has no real zeros.

If the discriminant is zero, then the parabola's highest or lowest point is on the *x*-axis, and it has a single real zero.

The highest or lowest point of the parabola is called the *vertex*. The coordinates of the vertex are given by the point $(-\frac{b}{2a}, -\frac{D}{4a})$. The roots of a quadratic function can be found with the quadratic formula, which is:

$$x = \frac{-b \pm \sqrt{b^2 - 4ac}}{2a}$$

A *rational function* is a function $f(x) = \frac{p(x)}{q(x)}$, where p and q are both polynomials. The domain of f will be all real numbers except the (real) roots of q.

At these roots, the graph of f will have a *vertical asymptote*, unless they are also roots of p. Here is an example to consider:

$$p(x) = p_n x^n + p_{n-1} x^{n-1} + p_{n-2} x^{n-2} + \cdots + p_1 x + p_0$$

$$q(x) = q_m x^m + q_{m-1} x^{m-1} + q_{m-2} x^{m-2} + \cdots + q_1 x + q_0$$

When the degree of p is less than the degree of q, there will be a horizontal asymptote of $y = 0$. If p and q have the same degree, there will be a horizontal asymptote of $y = \frac{p_n}{q_n}$. If the degree of p is exactly one greater than the degree of q, then f will have an oblique asymptote along the line:

$$y = \frac{p_n}{q_{n-1}} x + \frac{p_{n-1}}{q_{n-1}}$$

Not all equations in x and y can be written in the form $y = f(x)$. An equation can be written in such a form if it satisfies the *vertical line test*: no vertical line meets the graph of the equation at more than a single point. In this case, y is said to be a *function of x*. If a vertical line meets the graph in two places, then this equation cannot be written in the form $y = f(x)$.

The graph of a function $f(x)$ is the graph of the equation $y = f(x)$. Thus, it is the set of all pairs (x, y) where $y = f(x)$. In other words, it is all pairs $(x, f(x))$. The x-intercepts are called the *zeros* of the function. The y-intercept is given by $f(0)$.

If, for a given function f, the only way to get $f(a) = f(b)$ is for $a = b$, then f is *one-to-one*. Often, even if a function is not one-to-one on its entire domain, it is one-to-one by considering a restricted portion of the domain.

A function $f(x) = k$ for some number k is called a *constant function*. The graph of a constant function is a horizontal line.

The function $f(x) = x$ is called the *identity function*. The graph of the identity function is the diagonal line pointing to the upper right at 45 degrees, $y = x$.

Given two functions, $f(x)$ *and* $g(x)$, new functions can be formed by adding, subtracting, multiplying, or dividing the functions. Any algebraic combination of the two functions can be performed, including one function being the exponent of the other. If there are expressions for f and g, then the result can be found by performing the desired operation between the expressions. So, if $f(x) = x^2$ and $g(x) = 3x$, then:

$$f \cdot g(x) = x^2 \cdot 3x = 3x^3$$

Given two functions, $f(x)$ and $g(x)$, where the domain of g contains the range of f, the two functions can be combined together in a process called *composition*. The function—"g composed of f"—is written:

$$(g \circ f)(x) = g(f(x))$$

This requires the input of x into f, then taking that result and plugging it in to the function g.

If f is one-to-one, then there is also the option to find the function $f^{-1}(x)$, called the *inverse* of f. Algebraically, the inverse function can be found by writing y in place of $f(x)$, and then solving for x. The inverse function also makes this statement true:

$$f^{-1}(f(x)) = x$$

Computing the inverse of a function f entails the following procedure:

Given $f(x) = x^2$, with a domain of $x \geq 0$

$x = y^2$ is written down to find the inverse

The square root of both sides is determined to solve for y

Normally, this would mean $\pm\sqrt{x} = y$. However, the domain of f does not include the negative numbers, so the negative option needs to be eliminated.

The result is $y = \sqrt{x}$, so $f^{-1}(x) = \sqrt{x}$, with a domain of $x \geq 0$.

A function is called *monotone* if it is either always increasing or always decreasing. For example, the functions $f(x) = 3x$ and $f(x) = -x^5$ are monotone.

An *even function* looks the same when flipped over the y-axis:

$$f(x) = f(-x)$$

The following image shows a graphic representation of an even function.

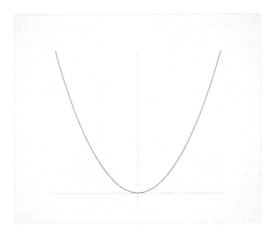

An *odd function* looks the same when flipped over the *y*-axis and then flipped over the *x*-axis:

$$f(x) = -f(-x)$$

The following image shows an example of an odd function.

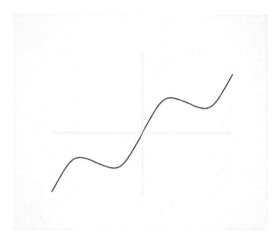

To evaluate functions, plug in the given value everywhere the variable appears in the expression for the function. For example, find $g(-2)$ where:

$$g(x) = 2x^2 - \frac{4}{x}$$

To complete the problem, plug in -2 in the following way:

$$g(-2) = 2(-2)^2 - \frac{4}{-2}$$

$$2 \cdot 4 + 2 = 8 + 2 = 10$$

For a Function that Models a Relationship Between Two Quantities, Interpret Key Features of Graphs and Tables

Scatterplots can be used to determine whether a correlation exists between two variables. The horizontal (*x*) axis represents the independent variable and the vertical (*y*) axis represents the dependent variable. If when graphed, the points model a linear, quadratic, or exponential relationship, then a correlation is said to exist. If so, a line of best-fit or curve of best-fit can be drawn through the points, with the points relatively close on either side. Writing the equation for the line or curve allows for predicting values for the variables. Suppose a scatterplot displays the value of an investment as a function of years after investing. By writing an equation for the line or curve and substituting a value for one variable into the equation, the corresponding value for the other variable can be calculated.

Linear Models

If the points of a scatterplot model a linear relationship, a line of best-fit is drawn through the points. If the line of best-fit has a positive slope (*y*-values increase as *x*-values increase), then the variables have a positive correlation. If the line of best-fit has a negative slope (*y*-values decrease as *x*-values increase), then a negative correlation exists. A positive or negative correlation can also be categorized as strong or weak, depending on how closely the points are grouped around the line of best-fit.

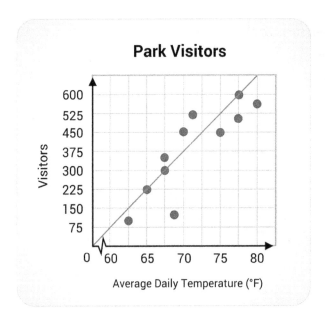

Given a line of best-fit, its equation can be written by identifying: the slope and y-intercept; a point and the slope; or two points on the line.

Quadratic Models

A quadratic function can be written in the form:

$$y = ax^2 + bx + c$$

The u-shaped graph of a quadratic function is called a parabola. The graph can either open up or open down (upside down u). The graph is symmetric about a vertical line, called the axis of symmetry. Corresponding points on the parabola are directly across from each other (same y-value) and are the same distance from the axis of symmetry (on either side). The axis of symmetry intersects the parabola at its vertex. The y-value of the vertex represents the minimum or maximum value of the function. If the graph opens up, the value of a in its equation is positive and the vertex represents the minimum of the function. If the graph opens down, the value of a in its equation is negative and the vertex represents the maximum of the function.

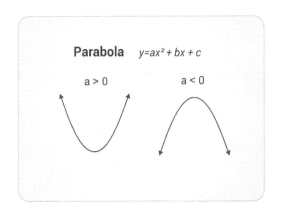

Given a curve of best-fit that models a quadratic relationship, the equation of the parabola can be written by identifying the vertex of the parabola and another point on the graph. The values for the vertex (h, k) and the point (x, y) should be substituted into the vertex form of a quadratic function, $y = a(x - h)^2 - k$, to determine the value of a. To write the equation of a quadratic function with a vertex of (4, 7) and containing the point (8, 3), the values for h, k, x, and y should be substituted into the vertex form of a quadratic function, resulting in:

$$3 = a(8 - 4)^2 + 7$$

Solving for a, yields $a = -\frac{1}{4}$. Therefore, the equation of the function can be written as:

$$y = -\frac{1}{4}(x - 4)^2 + 7$$

The vertex form can be manipulated in order to write the quadratic function in standard form.

Exponential Models

An exponential curve can be used as a curve of best-fit for a scatterplot. The general form for an exponential function is $y = ab^x$ where b must be a positive number and cannot equal 1. When the value of b is greater than 1, the function models exponential growth (as x increases, y increases). When the value of b is less than 1, the function models exponential decay (as x increases, y decreases). If a is positive, the graph consists of points above the x-axis; and if a is negative, the graph consists of points below the x-axis. An asymptote is a line that a graph approaches.

Exponential Curve

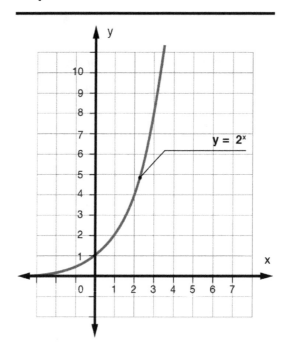

Given a curve of best-fit modeling an exponential function, its equation can be written by identifying two points on the curve. To write the equation of an exponential function containing the ordered pairs (2, 2) and (3, 4), the ordered pair (2, 2) should be substituted in the general form and solved for a:

$$2 = a \text{ x } b^2 \rightarrow a = \frac{2}{b^2}$$

The ordered pair (3, 4) and $\frac{2}{b^2}$ should be substituted in the general form and solved for b:

$$4 = \frac{2}{b^2} \text{ x } b^3 \rightarrow b = 2$$

Then, 2 should be substituted for b in the equation for a and then solved for a:

$$a = \frac{2}{2^2} \rightarrow a = \frac{1}{2}$$

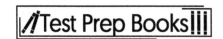

Knowing the values of *a* and *b*, the equation can be written as:

$$y = \frac{1}{2} \times 2^x$$

Solving Line Problems

Two lines are parallel if they have the same slope and different intercept. Two lines are perpendicular if the product of their slope equals -1. Parallel lines never intersect unless they are the same line, and perpendicular lines intersect at a right angle. If two lines aren't parallel, they must intersect at one point. Determining equations of lines based on properties of parallel and perpendicular lines appears in word problems. To find an equation of a line, both the slope and a point the line goes through are necessary. Therefore, if an equation of a line is needed that's parallel to a given line and runs through a specified point, the slope of the given line and the point are plugged into the point-slope form of an equation of a line. Secondly, if an equation of a line is needed that's perpendicular to a given line running through a specified point, the negative reciprocal of the slope of the given line and the point are plugged into the point-slope form. Also, if the point of intersection of two lines is known, that point will be used to solve the set of equations. Therefore, to solve a system of equations, the point of intersection must be found. If a set of two equations with two unknown variables has no solution, the lines are parallel.

Calculate and Interpret the Average Rate of Change of a Function

Rate of change for any line calculates the steepness of the line over a given interval. Rate of change is also known as the slope or rise/run. The rates of change for nonlinear functions vary depending on the interval being used for the function. The rate of change over one interval may be zero, while the next interval may have a positive rate of change. The equation plotted on the graph below, $y = x^2$, is a quadratic function and non-linear.

The average rate of change from points $(0, 0)$ to $(1, 1)$ is 1 because the vertical change is 1 over the horizontal change of 1. For the next interval, $(1, 1)$ to $(2, 4)$, the average rate of change is 3 because the slope is $\frac{3}{1}$.

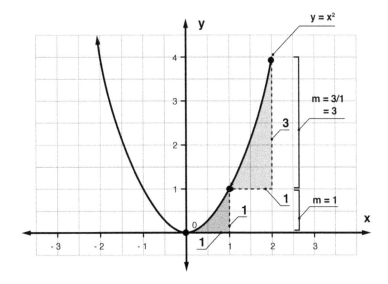

The rate of change for a linear function is constant and can be determined based on a few representations. One method is to place the equation in slope-intercept form:

$$y = mx + b$$

Thus, m is the slope, and b is the y-intercept. In the graph below, the equation is $y = x + 1$, where the slope is 1 and the y-intercept is 1. For every vertical change of 1 unit, there is a horizontal change of 1 unit. The x-intercept is -1, which is the point where the line crosses the x-axis.

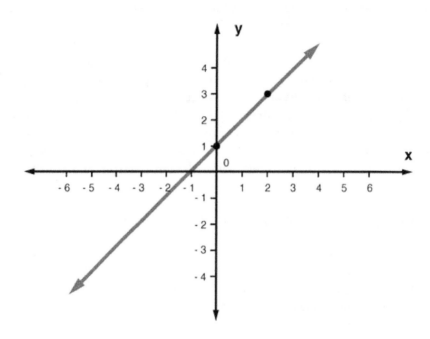

Graph Functions Expressed Symbolically

To graph relations and functions, the Cartesian plane is used. This means to think of the plane as being given a grid of squares, with one direction being the x-axis and the other direction the y-axis. Generally, the independent variable is placed along the horizontal axis, and the dependent variable is placed along the vertical axis. Any point on the plane can be specified by saying how far to go along the x-axis and how far along the y-axis with a pair of numbers (x, y). Specific values for these pairs can be given names such as $C = (-1, 3)$. Negative values mean to move left or down; positive values mean to move right or up. The point where the axes cross one another is called the *origin*. The origin has coordinates $(0, 0)$ and is usually called O when given a specific label. An illustration of the Cartesian plane, along with graphs of $(2, 1)$ and $(-1, -1)$, are below.

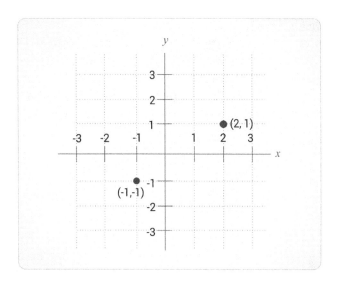

Relations also can be graphed by marking each point whose coordinates satisfy the relation. If the relation is a function, then there is only one value of y for any given value of x. This leads to the **vertical line test**: if a relation is graphed, then the relation is a function if any possible vertical line drawn anywhere along the graph would only touch the graph of the relation in no more than one place. Conversely, when graphing a function, then any possible vertical line drawn will not touch the graph of the function at any point or will touch the function at just one point. This test is made from the definition of a function, where each x-value must be mapped to one and only one y-value.

Understanding Connections Between Algebraic and Graphical Representations

The solution set to a linear equation in two variables can be represented visually by a line graphed on the coordinate plane. Every point on this line represents an ordered pair (x, y), which makes the equation true. The process for graphing a line depends on the form in which its equation is written: slope-intercept form or standard form.

Graphing a Line in Slope-Intercept Form

When an equation is written in slope-intercept form, $y = mx + b$, m represents the slope of the line and b represents the y-intercept. The y-intercept is the value of y when $x = 0$ and the point at which the graph of the line crosses the y-axis. The slope is the rate of change between the variables, expressed as a fraction. The fraction expresses the change in y compared to the change in x. If the slope is an integer, it should be written as a fraction with a denominator of 1. For example, 5 would be written as 5/1.

163

To graph a line given an equation in slope-intercept form, the *y*-intercept should first be plotted. For example, to graph $y = -\frac{2}{3}x + 7$, the *y*-intercept of 7 would be plotted on the *y*-axis (vertical axis) at the point (0, 7). Next, the slope would be used to determine a second point for the line. Note that all that is necessary to graph a line is two points on that line. The slope will indicate how to get from one point on the line to another. The slope expresses vertical change (*y*) compared to horizontal change (*x*) and therefore is sometimes referred to as $\frac{rise}{run}$. The numerator indicates the change in the *y* value (move up for positive integers and move down for negative integers), and the denominator indicates the change in the *x* value. For the previous example, using the slope of $-\frac{2}{3}$, from the first point at the *y*-intercept, the second point should be found by counting down 2 and to the right 3. This point would be located at (3, 5).

Graphing a Line in Standard Form

When an equation is written in standard form, $Ax + By = C$, it is easy to identify the *x*- and *y*-intercepts for the graph of the line. Just as the *y*-intercept is the point at which the line intercepts the *y*-axis, the *x*-intercept is the point at which the line intercepts the *x*-axis. At the *y*-intercept, $x = 0$; and at the *x*-intercept, $y = 0$. Given an equation in standard form, $x = 0$ should be used to find the *y*-intercept. Likewise, $y = 0$ should be used to find the *x*-intercept. For example, to graph $3x + 2y = 6$, 0 for *y* results in:

$$3x + 2(0) = 6$$

Solving for *y* yields $x = 2$; therefore, an ordered pair for the line is (2, 0). Substituting 0 for *x* results in:

$$3(0) + 2y = 6$$

Solving for *y* yields $y = 3$; therefore, an ordered pair for the line is (0, 3). The two ordered pairs (the *x*- and *y*-intercepts) can be plotted and a straight line through them can be constructed.

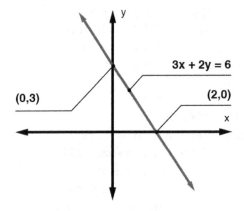

T - chart

x	y
0	3
2	0

Intercepts

x - intercept : (2,0)

y - intercept : (0,3)

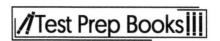

Writing the Equation of a Line Given its Graph

Given the graph of a line, its equation can be written in two ways. If the y-intercept is easily identified (is an integer), it and another point can be used to determine the slope. When determining $\frac{change\ in\ y}{change\ in\ x}$ from one point to another on the graph, the distance for $\frac{rise}{run}$ is being figured. The equation should be written in slope-intercept form, $y = mx + b$, with m representing the slope and b representing the y-intercept.

The equation of a line can also be written by identifying two points on the graph of the line. To do so, the slope is calculated and then the values are substituted for the slope and either of the ordered pairs into the point-slope form of an equation.

Vertical, Horizontal, Parallel, and Perpendicular Lines

For a vertical line, the value of x remains constant (for all ordered pairs (x, y) on the line, the value of x is the same); therefore, the equations for all vertical lines are written in the form $x = number$. For example, a vertical line that crosses the x-axis at -2 would have an equation of $x = -2$. For a horizontal line, the value of y remains constant; therefore, the equations for all horizontal lines are written in the form $y = number$.

Parallel lines extend in the same exact direction without ever meeting. Their equations have the same slopes and different y-intercepts. For example, given a line with an equation of $y = -3x + 2$, a parallel line would have a slope of -3 and a y-intercept of any value other than 2. Perpendicular lines intersect to form a right angle. Their equations have slopes that are opposite reciprocal (the sign is changed and the fraction is flipped; for example, $-\frac{2}{3}$ and $\frac{3}{2}$) and y-intercepts that may or may not be the same. For example, given a line with an equation of $y = \frac{1}{2}x + 7$, a perpendicular line would have a slope of $-\frac{2}{1}$ and any value for its y-intercept.

Use Properties of Exponents to Interpret Expressions for Exponential Functions

An *exponential function* is a function of the form $f(x) = b^x$, where b is a positive real number other than 1. In such a function, b is called the *base*.

The *domain* of an exponential function is all real numbers, and the *range* is all positive real numbers. There will always be a horizontal asymptote of $y = 0$ on one side. If b is greater than 1, then the graph will be increasing moving to the right. If b is less than 1, then the graph will be decreasing moving to the right. Exponential functions are one-to-one. The basic exponential function graph will go through the point (0,1).

Example
Solve $5^{x+1} = 25$.

Get the x out of the exponent by rewriting the equation $5^{x+1} = 5^2$ so that both sides have a base of 5.

Since the bases are the same, the exponents must be equal to each other.

This leaves $x + 1 = 2$ or $x = 1$.

To check the answer, the x-value of 1 can be substituted back into the original equation.

Solving Logarithmic and Exponential Functions

To solve an equation involving exponential expressions, the goal is to isolate the exponential expression. Once this process is completed, the logarithm—with the base equaling the base of the exponent of both sides—needs to be taken to get an expression for the variable. If the base is e, the natural log of both sides needs to be taken.

To solve an equation with logarithms, the given equation needs to be written in exponential form, using the fact that $\log_b y = x$ means $b^x = y$, and then solved for the given variable. Lastly, properties of logarithms can be used to simplify more than one logarithmic expression into one.

When working with logarithmic functions, it is important to remember the following properties. Each one can be derived from the definition of the logarithm as the inverse to an exponential function:

$$\log_b 1 = 0$$

$$\log_b b = 1$$

$$\log_b b^p = p$$

$$\log_b MN = \log_b M + \log_b N$$

$$\log_b \frac{M}{N} = \log_b M - \log_b N$$

$$\log_b M^p = p \log_b M$$

When solving equations involving exponentials and logarithms, the following fact should be used:

If f is a one-to-one function, $a = b$ is equivalent to $f(a) = f(b)$.

Using this, together with the fact that logarithms and exponentials are inverses, allows manipulations of the equations to isolate the variable.

Example
Solve $4 = \ln(x - 4)$.

Using the definition of a logarithm, the equation can be changed to:

$$e^4 = e^{\ln(x-4)}$$

The functions on the right side cancel with a result of:

$$e^4 = x - 4$$

This then gives:

$$x = 4 + e^4$$

Exponential Expressions

Exponential expressions can also be rewritten, just as quadratic equations. Properties of exponents must be understood. Multiplying two exponential expressions with the same base involves adding the exponents:

$$a^m a^n = a^{m+n}$$

Dividing two exponential expressions with the same base involves subtracting the exponents:

$$\frac{a^m}{a^n} = a^{m-n}$$

Raising an exponential expression to another exponent includes multiplying the exponents:

$$(a^m)^n = a^{mn}$$

The zero power always gives a value of 1: $a^0 = 1$. Raising either a product or a fraction to a power involves distributing that power:

$$(ab)^m = a^m b^m \text{ and } \left(\frac{a}{b}\right)^m = \frac{a^m}{b^m}$$

Finally, raising a number to a negative exponent is equivalent to the reciprocal including the positive exponent:

$$a^{-m} = \frac{1}{a^m}$$

Compare Properties of Two Functions Represented in a Different Ways

Mathematical functions such as polynomials, rational functions, radical functions, absolute value functions, and piecewise-defined functions can be utilized to approximate, or model, real-life phenomena. For example, a function can be built that approximates the average amount of snowfall on a given day of the year in Chicago. This example could be as simple as a polynomial. Modeling situations using such functions has limitations; the most significant issue is the error that exists between the exact amount and the approximate amount. Typically, the model will not give exact values as outputs. However, choosing the type of function that provides the best fit of the data will reduce this error. Technology can be used to model situations. For example, given a set of data, the data can be inputted into tools such as graphing calculators or spreadsheet software that output a function with a good fit. Some examples of polynomial modeling are linear, quadratic, and cubic regression.

Representing Exponential and Logarithmic Functions

The logarithmic function with base b is denoted $y = \log_b x$. Its base must be greater than 0 and not equal to 1, and the domain is all $x > 0$. The exponential function with base b is denoted $y = b^x$. Exponential and logarithmic functions with base b are inverses. By definition, if $y = \log_b x$, $x = b^y$. Because exponential and logarithmic functions are inverses, the graph of one is obtained by reflecting the other over the line $y = x$. A common base used is e, and in this case $y = e^x$ and its inverse $y = \log_e x$ is commonly written as the natural logarithmic function $y = \ln x$.

Here is the graph of both functions:

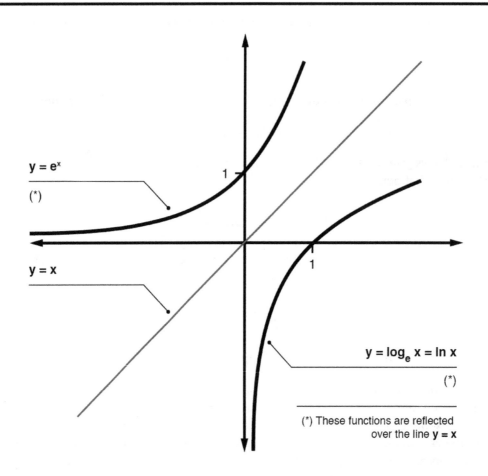

The Graphs of Exponential and Logarithmic Functions are Inverses

y = e^x

(*)

y = x

y = log_e x = ln x

(*)

(*) These functions are reflected over the line y = x

Graphing Functions

The x-intercept of the logarithmic function $y = \log_b x$ with any base is always the ordered pair $(1, 0)$. By the definition of inverse, the point $(0, 1)$ always lies on the exponential function $y = b^x$. This is true because any real number raised to the power of 0 equals 1. Therefore, the exponential function only has a y-intercept. The exponential function also has a horizontal asymptote of the x-axis as x approaches negative infinity. Because the graph is reflected over the line $y = x$, to obtain the graph of the logarithmic function, the asymptote is also reflected. Therefore, the logarithmic function has a one-sided vertical asymptote at $y = 0$. These asymptotes can be seen in the above graphs of $y = e^x$ and $y = \ln x$.

Solving Logarithmic and Exponential Functions

To solve an equation involving exponential expressions, the goal is to isolate the exponential expression. Once this process is completed, the logarithm—with the base equaling the base of the exponent of both

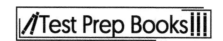

sides—needs to be taken to get an expression for the variable. If the base is e, the natural log of both sides needs to be taken.

To solve an equation with logarithms, the given equation needs to be written in exponential form, using the fact that $\log_b a = x$ means $b^x = y$, and then solved for the given variable. Lastly, properties of logarithms can be used to simplify more than one logarithmic expression into one.

Some equations involving exponential and logarithmic functions can be solved algebraically, or analytically. To solve an equation involving exponential functions, the goal is to isolate the exponential expression. Then, the logarithm of both sides is found in order to yield an expression for the variable. Laws of Logarithms will be helpful at this point.

To solve an equation with logarithms, the equation needs to be rewritten in exponential form. The definition that $\log_b x = y$ means $b^y = x$ needs to be used. Then, one needs to solve for the given variable. Properties of logarithms can be used to simplify multiple logarithmic expressions into one.

Other methods can be used to solve equations containing logarithmic and exponential functions. Graphs and graphing calculators can be used to see points of intersection. In a similar manner, tables can be used to find points of intersection. Also, numerical methods can be utilized to find approximate solutions.

Exponential Growth and Decay
Exponential growth and decay are important concepts in modeling real-world phenomena. The growth and decay formula is $A(t) = Pe^{rt}$, where the independent variable t represents temperature, P represents an initial quantity, r represents the rate of increase or decrease, and $A(t)$ represents the amount of the quantity at time t. If $r > 0$, the equation models exponential growth and a common application is population growth. If $r < 0$, the equation models exponential decay and a common application is radioactive decay. Exponential and logarithmic solving techniques are necessary to work with the growth and decay formula.

Logarithmic Scales
A logarithmic scale is a scale of measurement that uses the logarithm of the given units instead of the actual given units. Each tick mark on such a scale is the product of the previous tick mark multiplied by a number. The advantage of using such a scale is that if one is working with large measurements, this technique reduces the scale into manageable quantities that are easier to read. The Richter magnitude scale is the famous logarithmic scale used to measure the intensity of earthquakes, and the decibel scale is commonly used to measure sound level in electronics.

Using Exponential and Logarithmic Functions in Finance Problems
Modeling within finance also involves exponential and logarithmic functions. Compound interest results when the bank pays interest on the original amount of money – the principal – and the interest that has accrued. The compound interest equation is $A(t) = P\left(1 + \frac{r}{n}\right)^{nt}$, where P is the principal, r is the interest rate, n is the number of times per year the interest is compounded, and t is the time in years. The result, $A(t)$, is the final amount after t years. Mathematical problems of this type that are frequently encountered involve receiving all but one of these quantities and solving for the missing quantity. The solving process then involves employing properties of logarithmic and exponential functions. Interest can also be compounded continuously. This formula is given as:

$$A(t) = Pe^{rt}$$

If \$1,000 was compounded continuously at a rate of 2% for 4 years, the result would be:

$$A(4) = 1000e^{0.02 \cdot 4} = \$1,083$$

Rate of Change Proportional to the Current Quantity

Many quantities grow or decay as fast as exponential functions. Specifically, if such a quantity grows or decays at a rate proportional to the quantity itself, it shows exponential behavior. If a data set is given with such specific characteristics, the initial amount and an amount at a specific time, t, can be plugged into the exponential function $A(t) = Pe^{rt}$ for A and P. Using properties of exponents and logarithms, one can then solve for the rate, r. This solution yields enough information to have the entire model, which can allow for an estimation of the quantity at any time, t, and the ability to solve various problems using that model.

Write a Function that Describes a Relationship Between Two Quantities

Different types of functions behave in different ways. A function is defined to be increasing over a subset of its domain if for all $x_1 \geq x_2$ in that interval:

$$f(x_1) \geq f(x_2)$$

Also, a function is decreasing over an interval if for all $x_1 \geq x_2$ in that interval:

$$f(x_1) \leq f(x_2)$$

A point in which a function changes from increasing to decreasing can also be labeled as the *maximum value* of a function if it is the largest point the graph reaches on the y-axis. A point in which a function changes from decreasing to increasing can be labeled as the minimum value of a function if it is the smallest point the graph reaches on the y-axis. Maximum values are also known as *extreme values*. The graph of a continuous function does not have any breaks or jumps in the graph. This description is not true of all functions. A radical function, for example, $f(x) = \sqrt{x}$, has a restriction for the domain and range because there are no real negative inputs or outputs for this function. The domain can be stated as $x \geq 0$, and the range is $y \geq 0$.

A piecewise-defined function also has a different appearance on the graph. In the following function, there are three equations defined over different intervals. It is a function because there is only one y-value for each x-value, passing the Vertical Line Test. The domain is all real numbers less than or equal to 6. The range is all real numbers greater than zero. From left to right, the graph decreases to zero, then increases to almost 4, and then jumps to 6.

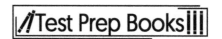

From input values greater than 2, the input decreases just below 8 to 4, and then stops.

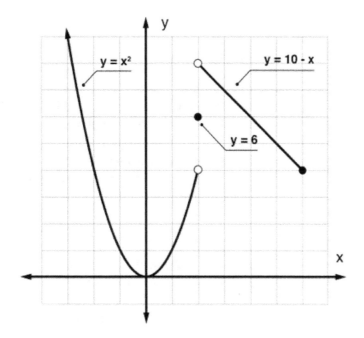

Logarithmic and exponential functions also have different behavior than other functions. These two types of functions are inverses of each other. The *inverse* of a function can be found by switching the place of x and y, and solving for y. When this is done for the exponential equation, $y = 2^x$, the function $y = \log_2 x$ is found. The general form of a *logarithmic function* is $y = \log_b x$, which says b raised to the y power equals x.

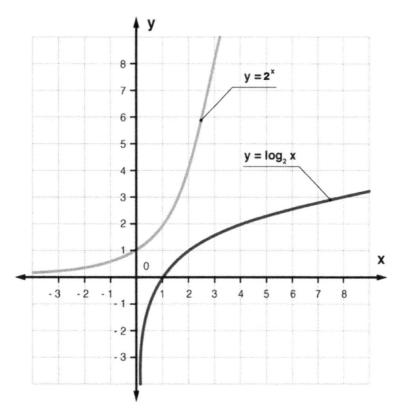

The thick black line on the graph above represents the logarithmic function:

$$y = \log_2 x$$

This curve passes through the point $(1, 0)$, just as all log functions do, because any value $b^0 = 1$. The graph of this logarithmic function starts very close to zero, but does not touch the y-axis. The output value will never be zero by the definition of logarithms. The thinner gray line seen above represents the exponential function $y = 2^x$. The behavior of this function is opposite the logarithmic function because the graph of an inverse function is the graph of the original function flipped over the line $y = x$. The curve passes through the point $(0, 1)$ because any number raised to the zero power is one. This curve also gets very close to the x-axis but never touches it because an exponential expression never has an output of zero. The x-axis on this graph is called a horizontal asymptote. An *asymptote* is a line that represents a boundary for a function. It shows a value that the function will get close to, but never reach.

Functions can also be described as being even, odd, or neither. If $f(-x) = f(x)$, the function is even. For example, the function $f(x) = x^2 - 2$ is even. Plugging in $x = 2$ yields an output of $y = 2$. After changing the input to $x = -2$, the output is still $y = 2$. The output is the same for opposite inputs. Another way to observe an even function is by the symmetry of the graph. If the graph is symmetrical about the axis, then the function is even. If the graph is symmetric about the origin, then the function is odd. Algebraically, if $f(-x) = -f(x)$, the function is odd.

Also, a function can be described as periodic if it repeats itself in regular intervals. Common periodic functions are trigonometric functions. For example, $y = \sin x$ is a periodic function with period 2π because it repeats itself every 2π units along the x-axis.

Building a Linear Function that Models a Linear Relationship Between Two Quantities

Linear relationships between two quantities can be expressed in two ways: function notation or as a linear equation with two variables. The relationship is referred to as linear because its graph is represented by a line. For a relationship to be linear, both variables must be raised to the first power only.

Function/Linear Equation Notation

A relation is a set of input and output values that can be written as ordered pairs. A function is a relation in which each input is paired with exactly one output. The domain of a function consists of all inputs, and the range consists of all outputs. Graphing the ordered pairs of a linear function produces a straight line. An example of a function would be $f(x) = 4x + 4$, read "f of x is equal to four times x plus four." In this example, the input would be x and the output would be f(x). Ordered pairs would be represented as (x, f(x)). To find the output for an input value of 3, 3 would be substituted for x into the function as follows: $f(3) = 4(3) + 4$, resulting in $f(3) = 16$. Therefore, the ordered pair:

$$(3, f(3)) = (3, 16)$$

Note f(x) is a function of x denoted by f. Functions of x could be named g(x), read "g of x"; p(x), read "p of x"; etc.

A linear function could also be written in the form of an equation with two variables. Typically, the variable *x* represents the inputs and the variable *y* represents the outputs. The variable *x* is considered the independent variable and *y* the dependent variable. The above function would be written as:

$$y = 4x + 4$$

Ordered pairs are written in the form (*x*, *y*).

Writing Linear Equations in Two Variables
When writing linear equations in two variables, the process depends on the information given. Questions will typically provide the slope of the line and its *y*-intercept, an ordered pair and the slope, or two ordered pairs.

Given the Slope and Y-Intercept
Linear equations are commonly written in slope-intercept form, $y = mx + b$, where *m* represents the slope of the line and *b* represents the *y*-intercept. The slope is the rate of change between the variables, usually expressed as a whole number or fraction. The *y*-intercept is the value of *y* when *x* = 0 (the point where the line intercepts the *y*-axis on a graph). Given the slope and *y*-intercept of a line, the values are substituted for *m* and *b* into the equation. A line with a slope of ½ and *y*-intercept of -2 would have an equation $y = ½x - 2$.

Given an Ordered Pair and the Slope
The point-slope form of a line, $y - y_1 = m(x - x_1)$, is used to write an equation when given an ordered pair (point on the equation's graph) for the function and its rate of change (slope of the line). The values for the slope, *m*, and the point (x_1, y_1) are substituted into the point-slope form to obtain the equation of the line. A line with a slope of 3 and an ordered pair (4, -2) would have an equation $y - (-2) = 3(x - 4)$. If a question specifies that the equation be written in slope-intercept form, the equation should be manipulated to isolate *y*:

Solve: $y - (-2) = 3(x - 4)$

Distribute: $y + 2 = 3x - 12$

Subtract 2 from both sides: $y = 3x - 14$

Given Two Ordered Pairs
Given two ordered pairs for a function, (x_1, y_1) and (x_2, y_2), it is possible to determine the rate of change between the variables (slope of the line). To calculate the slope of the line, m, the values for the ordered pairs should be substituted into the formula:

$$m = \frac{y_2 - y_1}{x_2 - x_1}$$

The expression is substituted to obtain a whole number or fraction for the slope. Once the slope is calculated, the slope and either of the ordered pairs should be substituted into the point-slope form to obtain the equation of the line.

Building a Function for a Given Situation
Functions can be built out of the context of a situation. For example, the relationship between the money paid for a gym membership and the months that someone has been a member can be described

through a function. If the one-time membership fee is \$40 and the monthly fee is \$30, then the function can be written:

$$f(x) = 30x + 40$$

The x-value represents the number of months the person has been part of the gym, while the output is the total money paid for the membership. The table below shows this relationship. It is a representation of the function because the initial cost is \$40 and the cost increases each month by \$30.

x (months)	y (money paid to gym)
0	40
1	70
2	100
3	130

Functions can also be built from existing functions. For example, a given function $f(x)$ can be transformed by adding a constant, multiplying by a constant, or changing the input value by a constant. The new function $g(x) = f(x) + k$ represents a vertical shift of the original function. In $f(x) = 3x - 2$, a vertical shift 4 units up would be:

$$g(x) = 3x - 2 + 4 = 3x + 2$$

Multiplying the function times a constant k represents a vertical stretch, based on whether the constant is greater than or less than 1. The function

$$g(x) = kf(x)$$

$$4(3x - 2) = 12x - 8$$

represents a stretch. Changing the input x by a constant forms the function:

$$g(x) = f(x + k)$$

$$3(x + 4) - 2 = 3x + 12 - 2$$

$$3x + 10$$

and this represents a horizontal shift to the left 4 units. If $(x - 4)$ was plugged into the function, it would represent a vertical shift.

A composition function can also be formed by plugging one function into another. In function notation, this is written:

$$(f \circ g)(x) = f(g(x))$$

For two functions $f(x) = x^2$ and $g(x) = x - 3$, the composition function becomes:

$$f(g(x)) = (x - 3)^2$$

$$x^2 - 6x + 9$$

The composition of functions can also be used to verify if two functions are inverses of each other. Given the two functions $f(x) = 2x + 5$ and $g(x) = \frac{x-5}{2}$, the composition function can be found $(f \circ g)(x)$. Solving this equation yields:

$$f(g(x)) = 2\left(\frac{x-5}{2}\right) + 5$$

$$x - 5 + 5 = x$$

It also is true that"

$$g(f(x)) = x$$

Since the composition of these two functions gives a simplified answer of x, this verifies that $f(x)$ and $g(x)$ are inverse functions. The domain of $f(g(x))$ is the set of all x-values in the domain of $g(x)$ such that $g(x)$ is in the domain of $f(x)$. Basically, both $f(g(x))$ and $g(x)$ have to be defined.

To build an inverse of a function, $f(x)$ needs to be replaced with y, and the x and y values need to be switched. Then, the equation can be solved for y. For example, given the equation $y = e^{2x}$, the inverse can be found by rewriting the equation $x = e^{2y}$. The natural logarithm of both sides is taken down, and the exponent is brought down to form the equation:

$$\ln(x) = \ln(e)\, 2y$$

ln (e)=1, which yields the equation $\ln(x) = 2y$. Dividing both sides by 2 yields the inverse equation

$$\frac{\ln(x)}{2} = y = f^{-1}(x)$$

The domain of an inverse function is the range of the original function, and the range of an inverse function is the domain of the original function. Therefore, an ordered pair (x, y) on either a graph or a table corresponding to $f(x)$ means that the ordered pair (y, x) exists on the graph of $f^{-1}(x)$. Basically, if $f(x) = y$, then $f^{-1}(y) = x$. For a function to have an inverse, it must be one-to-one. That means it must pass the *Horizontal Line Test*, and if any horizontal line passes through the graph of the function twice, a function is not one-to-one. The domain of a function that is not one-to-one can be restricted to an interval in which the function is one-to-one, to be able to define an inverse function.

Functions can also be formed from combinations of existing functions.

Given $f(x)$ and $g(x)$, the following can be built:

$$f + g$$

$$f - g$$

$$fg$$

$$\frac{f}{g}$$

The domains of $f + g, f - g,$ and fg are the intersection of the domains of f and g. The domain of $\frac{f}{g}$ is the same set, excluding those values that make $g(x) = 0$.

For example, if:

$$f(x) = 2x + 3$$

$$g(x) = x + 1$$

then

$$\frac{f}{g} = \frac{2x + 3}{x + 1}$$

Its domain is all real numbers except -1.

Recognize Situations in Which a Quantity Grows or Decays by a Constant Percent Rate Per Unit Interval Relative to Another

Linear relationships describe the way two quantities change with respect to each other. The relationship is defined as linear because a line is produced if all the sets of corresponding values are graphed on a coordinate grid. When expressing the linear relationship as an equation, the equation is often written in the form $y = mx + b$ (slope-intercept form) where m and b are numerical values and x and y are variables (for example, $y = 5x + 10$). Given a linear equation and the value of either variable (x or y), the value of the other variable can be determined.

Suppose a teacher is grading a test containing 20 questions with 5 points given for each correct answer, adding a curve of 10 points to each test. This linear relationship can be expressed as the equation $y = 5x + 10$ where x represents the number of correct answers and y represents the test score. To determine the score of a test with a given number of correct answers, the number of correct answers is substituted into the equation for x and evaluated. For example, for 10 correct answers, 10 is substituted for x:

$$y = 5(10) + 10 \rightarrow y = 60$$

Therefore, 10 correct answers will result in a score of 60. The number of correct answers needed to obtain a certain score can also be determined. To determine the number of correct answers needed to score a 90, 90 is substituted for y in the equation (y represents the test score) and solved:

$$90 = 5x + 10$$

$$80 = 5x \rightarrow 16 = x$$

Therefore, 16 correct answers are needed to score a 90.

Linear relationships may be represented by a table of 2 corresponding values. Certain tables may determine the relationship between the values and predict other corresponding sets. Consider the table below, which displays the money in a checking account that charges a monthly fee:

Month	0	1	2	3	4
Balance	$210	$195	$180	$165	$150

An examination of the values reveals that the account loses $15 every month (the month increases by one and the balance decreases by 15). This information can be used to predict future values. To determine what the value will be in month 6, the pattern can be continued, and it can be concluded that the balance will be $120. To determine which month the balance will be $0, $210 is divided by $15 (since the balance decreases $15 every month), resulting in month 14.

Similar to a table, a graph can display corresponding values of a linear relationship.

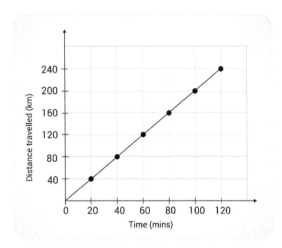

The graph above represents the relationship between distance traveled and time. To find the distance traveled in 80 minutes, the mark for 80 minutes is located at the bottom of the graph. By following this mark directly up on the graph, the corresponding point for 80 minutes is directly across from the 160-kilometer mark. This information indicates that the distance travelled in 80 minutes is 160 kilometers. To predict information not displayed on the graph, the way in which the variables change with respect to one another is determined. In this case, distance increases by 40 kilometers as time increases by 20 minutes. This information can be used to continue the data in the graph or convert the values to a table.

Interpret the Parameters in a Linear or Exponential Function

Three common functions used to model different relationships between quantities are linear, quadratic, and exponential functions. Linear functions are the simplest of the three, and the independent variable x has an exponent of 1. Written in the most common form, $y = mx + b$, the coefficient of x tells how fast the function grows at a constant rate, and the b-value tells the starting point. A quadratic function has an exponent of 2 on the independent variable x. Standard form for this type of function is $y = ax^2 + bx + c$, and the graph is a parabola. These type functions grow at a changing rate. An exponential function has an independent variable in the exponent $y = ab^x$. The graph of these types of functions is described as *growth* or *decay*, based on whether the base, b, is greater than or less than 1. These functions are different from quadratic functions because the base stays constant. A common base is base e.

The following three functions model a linear, quadratic, and exponential function respectively: $y = 2x$, $y = x^2$, and $y = 2^x$. Their graphs are shown below. The first graph, modeling the linear function, shows that the growth is constant over each interval. With a horizontal change of 1, the vertical change is 2. It models a constant positive growth. The second graph shows the quadratic function, which is a curve that is symmetric across the y-axis. The growth is not constant, but the change is mirrored over the axis. The last graph models the exponential function, where the horizontal change of 1 yields a vertical change that increases more and more. The exponential graph gets very close to the x-axis, but never touches it, meaning there is an asymptote there. The y-value can never be zero because the base of 2 can never be raised to an input value that yields an output of zero.

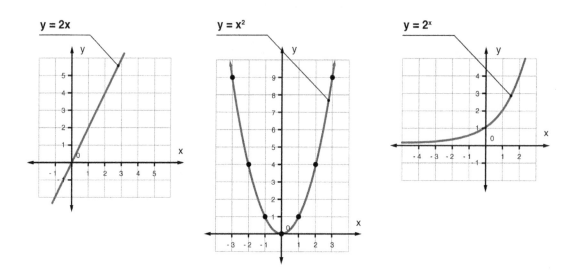

The three tables below show specific values for three types of functions. The third column in each table shows the change in the y-values for each interval. The first table shows a constant change of 2 for each equal interval, which matches the slope in the equation $y = 2x$. The second table shows an increasing change, but it also has a pattern. The increase is changing by 2 more each time, so the change is quadratic. The third table shows the change as factors of the base, 2. It shows a continuing pattern of factors of the base.

y = 2x		
x	y	$\triangle y$
1	2	
2	4	2
3	6	2
4	8	2
5	10	2

y = x²		
x	y	$\triangle y$
1	1	
2	4	3
3	9	5
4	16	7
5	25	9

y = 2ˣ		
x	y	$\triangle y$
1	2	
2	4	2
3	8	4
4	16	8
5	32	16

Given a table of values, the type of function can be determined by observing the change in y over equal intervals. For example, the tables below model two functions. The changes in interval for the x-values is 1 for both tables. For the first table, the y-values increase by 5 for each interval. Since the change is constant, the situation can be described as a linear function. The equation would be $y = 5x + 3$. For the second table, the change for y is 5, 20, 100, and 500, respectively. The increases are multiples of 5, meaning the situation can be modeled by an exponential function. The equation $y = 5^x + 3$ models this situation.

x	y
0	3
1	8
2	13
3	18
4	23

x	y
0	3
1	8
2	28
3	128
4	628

Quadratic equations can be used to model real-world area problems. For example, a farmer may have a rectangular field that he needs to sow with seed. The field has length $x + 8$ and width $2x$. The formula for area should be used: $A = lw$. Therefore:

$$A = (x + 8) * 2x = 2x^2 + 16x$$

The possible values for the length and width can be shown in a table, with input x and output A. If the equation was graphed, the possible area values can be seen on the y-axis for given x-values.

Exponential growth and decay can be found in real-world situations. For example, if a piece of notebook paper is folded 25 times, the thickness of the paper can be found. To model this situation, a table can be used. The initial point is one-fold, which yields a thickness of 2 papers. For the second fold, the thickness is 4. Since the thickness doubles each time, the table below shows the thickness for the next few folds. Notice the thickness changes by the same factor each time. Since this change for a constant interval of folds is a factor of 2, the function is exponential. The equation for this is $y = 2^x$. For twenty-five folds, the thickness would be 33,554,432 papers.

x (folds)	y (paper thickness)
0	1
1	2
2	4
3	8
4	16
5	32

One exponential formula that is commonly used is the *interest formula*:

$$A = Pe^{rt}$$

In this formula, interest is compounded continuously. A is the value of the investment after the time, t, in years. P is the initial amount of the investment, r is the interest rate, and e is the constant equal to approximately 2.718. Given an initial amount of $200 and a time of 3 years, if interest is compounded continuously at a rate of 6%, the total investment value can be found by plugging each value into the

Test Prep Books

formula. The invested value at the end is $239.44. In more complex problems, the final investment may be given, and the rate may be the unknown. In this case, the formula becomes:

$$239.44 = 200e^{r3}$$

Solving for r requires isolating the exponential expression on one side by dividing by 200, yielding the equation:

$$1.20 = e^{r3}$$

Taking the natural log of both sides results in:

$$\ln(1.2) = r3$$

Using a calculator to evaluate the logarithmic expression:

$$r = 0.06 = 6\%$$

When working with logarithms and exponential expressions, it is important to remember the relationship between the two. In general, the logarithmic form is $y = log_b x$ for an exponential form $b^y = x$. Logarithms and exponential functions are inverses of each other.

Interpreting Variables and Constants in Expressions for Linear Functions in the Context Presented

Linear functions, also written as linear equations in two variables, can be written to model real-world scenarios. Questions on this material will provide information about a scenario and then request a linear equation to represent the scenario. The algebraic process for writing the equation will depend on the given information. The key to writing linear models is to decipher the information given to determine what it represents in the context of a linear equation (variables, slope, ordered pairs, etc.).

Identifying Variables for Linear Models

The first step to writing a linear model is to identify what the variables represent. A variable represents an unknown quantity, and in the case of a linear equation, a specific relationship exists between the two variables (usually *x* and *y*). Within a given scenario, the variables are the two quantities that are changing. The variable *x* is considered the independent variable and represents the inputs of a function. The variable *y* is considered the dependent variable and represents the outputs of a function. For example, if a scenario describes distance traveled and time traveled, distance would be represented by *y* and time represented by *x*. The distance traveled depends on the time spent traveling (time is independent). If a scenario describes the cost of a cab ride and the distance traveled, the cost would be represented by *y* and the distance represented by *x*. The cost of a cab ride depends on the distance traveled.

Identifying the Slope and Y-Intercept for Linear Models

The slope of the graph of a line represents the rate of change between the variables of an equation. In the context of a real-world scenario, the slope will tell the way in which the unknown quantities (variables) change with respect to each other. A scenario involving distance and time might state that someone is traveling at a rate of 45 miles per hour. The slope of the linear model would be 45. A scenario involving the cost of a cab ride and distance traveled might state that the person is charged $3 for each mile. The slope of the linear model would be 3.

The *y*-intercept of a linear function is the value of *y* when $x = 0$ (the point where the line intercepts the *y*-axis on the graph of the equation). It is sometimes helpful to think of this as a "starting point" for a linear function. Suppose for the scenario about the cab ride that the person is told that the cab company charges a flat fee of $5 plus $3 for each mile. Before traveling any distance ($x = 0$), the cost is $5. The *y*-intercept for the linear model would be 5.

Identifying Ordered Pairs for Linear Models

A linear equation with two variables can be written given a point (ordered pair) and the slope or given two points on a line. An ordered pair gives a set of corresponding values for the two variables (*x* and *y*). As an example, for a scenario involving distance and time, it is given that the person traveled 112.5 miles in 2 ½ hours. Knowing that *x* represents time and *y* represents distance, this information can be written as the ordered pair (2.5, 112.5).

Statistics and Probability

Represent Data with Plots on the Real Number Line

Interpretation of Tables, Charts, and Graphs

Data can be represented in many ways. It is important to be able to organize the data into categories that could be represented using one of these methods. Equally important is the ability to read these types of diagrams and interpret their meaning.

Data in Tables

One of the most common ways to express data is in a table. The primary reason for plugging data into a table is to make interpretation more convenient. It's much easier to look at the table than to analyze results in a narrative paragraph. When analyzing a table, pay close attention to the title, variables, and data.

Let's analyze a theoretical antibiotic study. The study has 6 groups, named A through F, and each group receives a different dose of medicine. The results of the study are listed in the table below.

Results of Antibiotic Studies		
Group	Dosage of Antibiotics in milligrams (mg)	Efficacy (% of participants cured)
A	0 mg	20%
B	20 mg	40%
C	40 mg	75%
D	60 mg	95%
E	80 mg	100%
F	100 mg	100%

Tables generally list the title immediately above the data. The title should succinctly explain what is listed below. Here, "Results of Antibiotic Studies" informs the audience that the data pertains to the results of scientific study on antibiotics.

Identifying the variables at play is one of the most important parts of interpreting data. Remember, the independent variable is intentionally altered, and its change is independent of the other variables. Here, the dosage of antibiotics administered to the different groups is the independent variable. The study is intentionally manipulating the strength of the medicine to study the related results. Efficacy is the dependent variable since its results *depend* on a different variable, the dose of antibiotics. Generally, the independent variable will be listed before the dependent variable in tables.

Also play close attention to the variables' labels. Here, the dose is expressed in milligrams (mg) and efficacy in percentages (%). Keep an eye out for questions referencing data in a different unit measurement, as discussed in the next topic, or questions asking for a raw number when only the percentage is listed.

Now that the nature of the study and variables at play have been identified, the data itself needs be interpreted. Group A did not receive any of the medicine. As discussed earlier, Group A is the control, as it reflects the amount of people cured in the same timeframe without medicine. It's important to see that efficacy positively correlates with the dosage of medicine. A question using this study might ask for the lowest dose of antibiotics to achieve 100% efficacy. Although Group E and Group F both achieve 100% efficacy, it's important to note that Group E reaches 100% with a lower dose.

Data in Graphs

Graphs provide a visual representation of data. The variables are placed on the two axes. The bottom of the graph is referred to as the horizontal axis or X-axis. The left-hand side of the graph is known as the vertical axis or Y-axis. Typically, the independent variable is placed on the X-axis, and the dependent variable is located on the Y-axis. Sometimes, the X-axis is a timeline, and the dependent variables for different trials or groups have been measured throughout points in time; time is still an independent variable, but is not always immediately thought of as the independent variable being studied.

The most common types of graphs are the bar graph and the line graph.

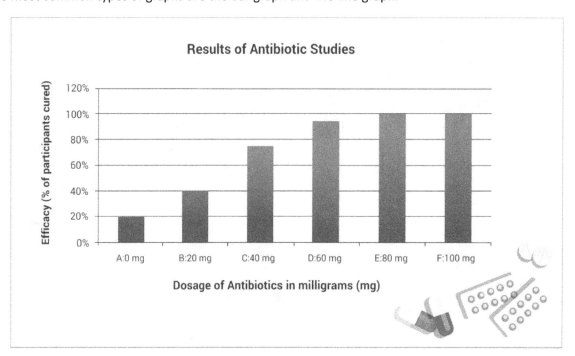

The *bar graph* above expresses the data from the table entitled "Results of Antibiotic Studies." To interpret the data for each group in the study, look at the top of their bars and read the corresponding efficacy on the Y-axis.

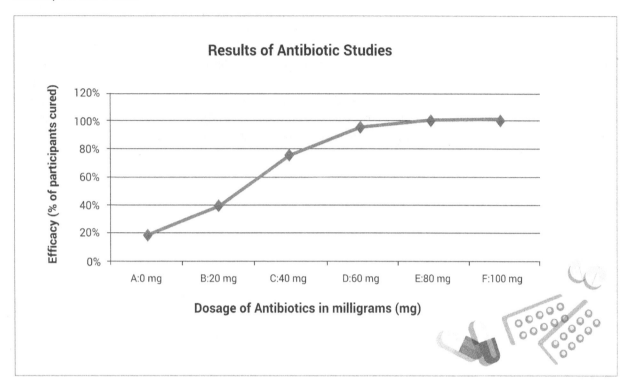

Here, the same data is expressed on a *line graph*. The points on the line correspond with each data entry. Reading the data on the line graph works like the bar graph. The data trend is measured by the slope of the line.

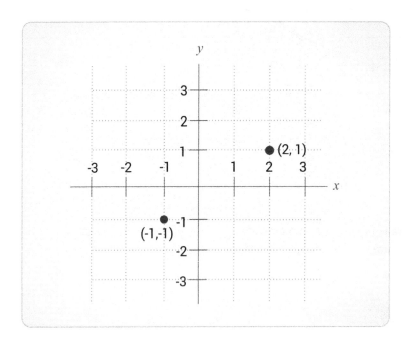 Test Prep Books!!!

Data in Other Charts

Chart is a broad term that refers to a variety of ways to represent data.

To graph relations, the *Cartesian plane* is used. This means to think of the plane as being given a grid of squares, with one direction being the *x*-axis and the other direction the *y*-axis. Generally, the independent variable is placed along the horizontal axis, and the dependent variable is placed along the vertical axis. Any point on the plane can be specified by saying how far to go along the *x*-axis and how far along the *y*-axis with a pair of numbers (x, y). Specific values for these pairs can be given names such as $C = (-1, 3)$. Negative values mean to move left or down; positive values mean to move right or up. The point where the axes cross one another is called the *origin*. The origin has coordinates $(0, 0)$ and is usually called *O* when given a specific label. An illustration of the Cartesian plane, along with graphs of $(2, 1)$ and $(-1, -1)$, are below.

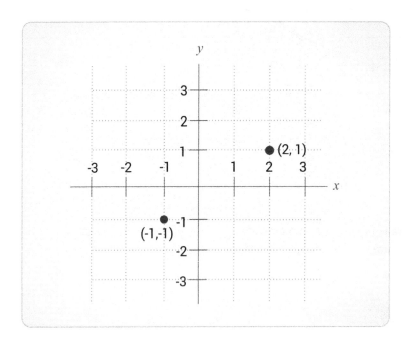

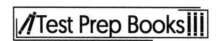

A *line plot* is a diagram that shows quantity of data along a number line. It is a quick way to record data in a structure similar to a bar graph without needing to do the required shading of a bar graph. Here is an example of a line plot:

A *tally chart* is a diagram in which tally marks are utilized to represent data. Tally marks are a means of showing a quantity of objects within a specific classification. Here is an example of a tally chart:

Number of days with rain	Number of weeks
0			
1	⊔⊔⊤
2	⊔⊔⊤	
3	⊔⊔⊤ ||||
4	⊔⊔⊤ ⊔⊔⊤ ⊔⊔⊤
5	⊔⊔⊤
6	⊔⊔⊤	
7					

Data is often recorded using fractions, such as half a mile, and understanding fractions is critical because of their popular use in real-world applications. Also, it is extremely important to label values with their units when using data. For example, regarding length, the number 2 is meaningless unless it is attached to a unit. Writing 2 cm shows that the number refers to the length of an object.

A *picture graph* is a diagram that shows pictorial representation of data being discussed. The symbols used can represent a certain number of objects. Notice how each fruit symbol in the following graph represents a count of two fruits. One drawback of picture graphs is that they can be less accurate if each symbol represents a large number. For example, if each banana symbol represented ten bananas, and

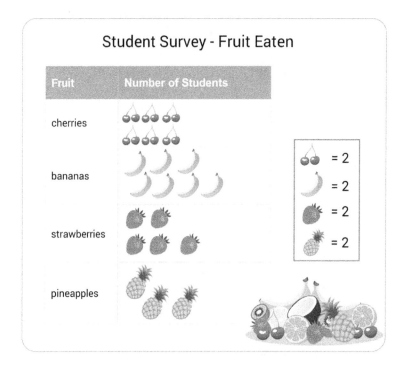

students consumed 22 bananas, it may be challenging to draw and interpret two and one-fifth bananas as a frequency count of 22.

A circle graph, also called a pie chart, shows categorical data with each category representing a percentage of the whole data set. To make a circle graph, the percent of the data set for each category must be determined. To do so, the frequency of the category is divided by the total number of data points and converted to a percent. For example, if 80 people were asked what their favorite sport is and 20 responded basketball, basketball makes up 25% of the data:

$$\frac{20}{80} = 0.25 = 25\%$$

Each category in a data set is represented by a *slice* of the circle proportionate to its percentage of the whole.

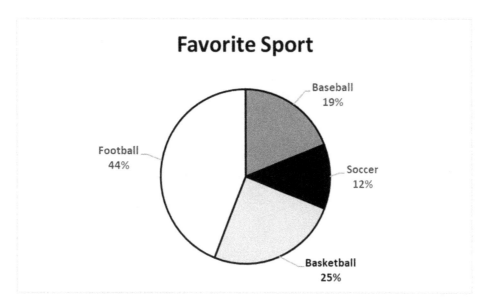

Favorite Sport

A scatter plot displays the relationship between two variables. Values for the independent variable, typically denoted by *x*, are paired with values for the dependent variable, typically denoted by *y*. Each set of corresponding values are written as an ordered pair (*x*, *y*). To construct the graph, a coordinate grid is labeled with the *x*-axis representing the independent variable and the *y*-axis representing the dependent variable. Each ordered pair is graphed.

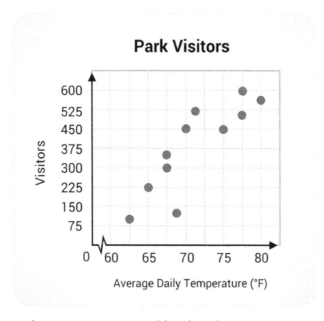

Park Visitors

Like a scatter plot, a line graph compares two variables that change continuously, typically over time. Paired data values (ordered pair) are plotted on a coordinate grid with the *x*- and *y*-axis representing the two variables. A line is drawn from each point to the next, going from left to right. A double line graph

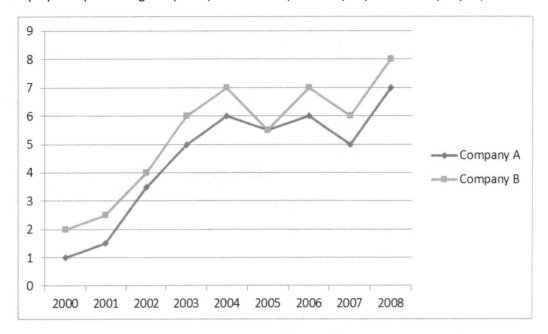
simply displays two sets of data that contain values for the same two variables. The double line graph below displays the profit for given years (two variables) for Company A and Company B (two data sets).

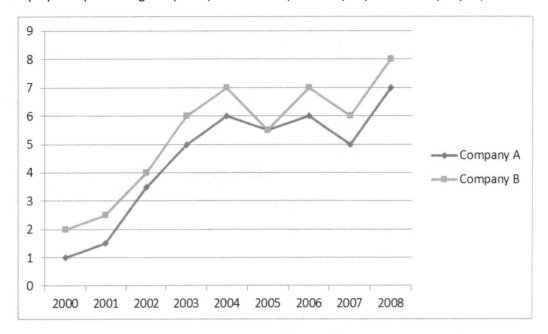

Choosing the appropriate graph to display a data set depends on what type of data is included in the set and what information must be shown.

Scatter plots and line graphs can be used to display data consisting of two variables. Examples include height and weight, or distance and time. A correlation between the variables is determined by examining the points on the graph. Line graphs are used if each value for one variable pairs with a distinct value for the other variable. Line graphs show relationships between variables.

Interpreting Competing Data

Be careful of questions with competing studies. These questions will ask to interpret which of two studies shows the greater amount or the higher rate of change between two results.

Here's an example. A research facility runs studies on two different antibiotics: Drug A and Drug B. The Drug A study includes 1,000 participants and cures 600 people. The Drug B study includes 200 participants and cures 150 people. Which drug is more successful?

The first step is to determine the percentage of each drug's rate of success. Drug A was successful in curing 60% of participants, while Drug B achieved a 75% success rate. Thus, Drug B is more successful based on these studies, even though it cured fewer people.

Sample size and experiment consistency should also be considered when answering questions based on competing studies. Is one study significantly larger than the other? In the antibiotics example, the Drug A study is five times larger than Drug B. Thus, Drug B's higher efficacy (desired result) could be a result of the smaller sample size, rather than the quality of drug.

Consistency between studies is directly related to sample size. Let's say the research facility elects to conduct more studies on Drug B. In the next study, there are 400 participants, and 200 are cured. The success rate of the second study is 50%. The results are clearly inconsistent with the first study, which means more testing is needed to determine the drug's efficacy. A hallmark of mathematical or scientific

research is repeatability. Studies should be consistent and repeatable, with an appropriately large sample size, before drawing extensive conclusions.

Histograms and Box Plots

To construct a histogram, the range of the data points is divided into equal intervals. The frequency for each interval is then determined, which reveals how many points fall into each interval. A graph is constructed with the vertical axis representing the frequency and the horizontal axis representing the intervals. The lower value of each interval should be labeled along the horizontal axis. Finally, for each interval, a bar is drawn from the lower value of each interval to the lower value of the next interval with a height equal to the frequency of the interval. Because of the intervals, histograms do not have any gaps between bars along the horizontal axis.

To construct a box (or box-and-whisker) plot, the five-number summary for the data set is calculated as follows: the second quartile (Q_2) is the median of the set. The first quartile (Q_1) is the median of the values below Q_2. The third quartile (Q_3) is the median of the values above Q_2. The upper extreme is the highest value in the data set if it is not an outlier (greater than 1.5 times the interquartile range Q_3 - Q_1). The lower extreme is the least value in the data set if it is not an outlier (more than 1.5 times lower than the interquartile range). To construct the box-and-whisker plot, each value is plotted on a number line, along with any outliers. The *box* consists of Q_1 and Q_3 as its *top* and *bottom* and Q_2 as the dividing line inside the box. The *whiskers* extend from the lower extreme to Q_1 and from Q_3 to the upper extreme.

Box Plot

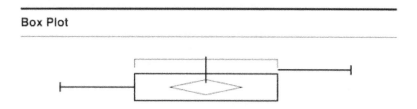

Choice of Graphs to Display Data

Choosing the appropriate graph to display a data set depends on what type of data is included in the set and what information must be displayed. Histograms and box plots can be used for data sets consisting of individual values across a wide range. Examples include test scores and incomes. Histograms and box plots will indicate the center, spread, range, and outliers of a data set. A histogram will show the shape of the data set, while a box plot will divide the set into quartiles (25% increments), allowing for comparison between a given value and the entire set.

Scatter plots and line graphs can be used to display data consisting of two variables. Examples include height and weight, or distance and time. A correlation between the variables is determined by examining the points on the graph. Line graphs are used if each value for one variable pairs with a distinct value for the other variable. Line graphs show relationships between variables.

Line plots, bar graphs, and circle graphs are all used to display categorical data, such as surveys. Line plots and bar graphs both indicate the frequency of each category within the data set. A line plot is used when the categories consist of numerical values. For example, the number of hours of TV watched by individuals is displayed on a line plot. A bar graph is used when the categories consists of words. For example, the favorite ice cream of individuals is displayed with a bar graph. A circle graph can be used to display either type of categorical data. However, unlike line plots and bar graphs, a circle graph does not indicate the frequency of each category. Instead, the circle graph represents each category as its percentage of the whole data set.

Interpret Differences in Shape, Center, and Spread in the Context of the Data Sets

A set of data can be described in terms of its center, spread, shape and any unusual features. The center of a data set can be measured by its mean, median, or mode. The spread of a data set refers to how far the data points are from the center (mean or median). The spread can be measured by the range or the quartiles and interquartile range. A data set with data points clustered around the center will have a small spread. A data set covering a wide range will have a large spread.

When a data set is displayed as a histogram or frequency distribution plot, the shape indicates if a sample is normally distributed, symmetrical, or has measures of skewness or kurtosis. When graphed, a data set with a normal distribution will resemble a bell curve.

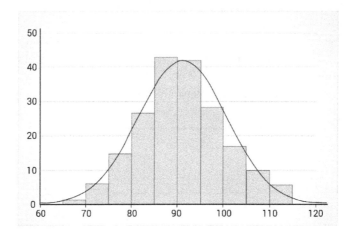

If the data set is symmetrical, each half of the graph when divided at the center is a mirror image of the other. If the graph has fewer data points to the right, the data is skewed right. If it has fewer data points to the left, the data is skewed left.

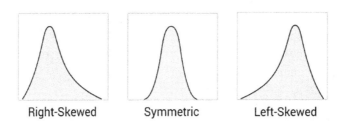

Right-Skewed   Symmetric   Left-Skewed

Kurtosis is a measure of whether the data is heavy-tailed with a high number of outliers, or light-tailed with a low number of outliers.

A description of a data set should include any unusual features such as gaps or outliers. A gap is a span within the range of the data set containing no data points. An outlier is a data point with a value either extremely large or extremely small when compared to the other values in the set.

Normal Distribution

A *normal distribution* of data follows the shape of a bell curve. In a normal distribution, the data set's median, mean, and mode are equal. Therefore, 50 percent of its values are less than the mean and 50 percent are greater than the mean. Data sets that follow this shape can be generalized using normal distributions. Normal distributions are described as *frequency distributions* in which the data set is plotted as percentages rather than true data points. A *relative frequency distribution* is one where the y-axis is between zero and 1, which is the same as 0% to 100%. Within a standard deviation, 68 percent of the values are within 1 standard deviation of the mean, 95 percent of the values are within 2 standard deviations of the mean, and 99.7 percent of the values are within 3 standard deviations of the mean. The number of standard deviations that a data point falls from the mean is called the *z-score*. The formula for the z-score is $Z = \frac{x-\mu}{\sigma}$, where μ is the mean, σ is the standard deviation, and x is the data point. This formula is used to fit any data set that resembles a normal distribution to a standard normal distribution, in a process known as *standardizing*. Here is a normal distribution with labelled z-scores:

Normal Distribution with Labelled Z-Scores

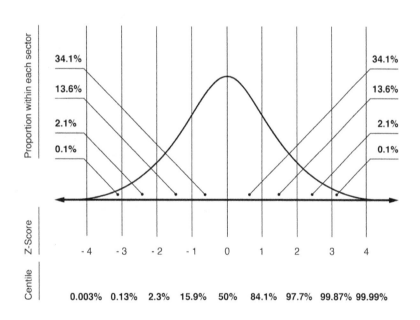

Population percentages can be estimated using normal distributions. For example, the probability that a data point will be less than the mean, or that the z-score will be less than 0, is 50%. Similarly, the probability that a data point will be within 1 standard deviation of the mean, or that the z-score will be between -1 and 1, is about 68.2%. When using a table, the left column states how many standard deviations (to one decimal place) away from the mean the point is, and the row heading states the second decimal place. The entries in the table corresponding to each column and row give the probability, which is equal to the area.

Measures of Center and Range

The center of a set of data (statistical values) can be represented by its mean, median, or mode. These are sometimes referred to as measures of central tendency. The mean is the average of the data set. The mean can be calculated by adding the data values and dividing by the sample size (the number of data points). Suppose a student has test scores of 93, 84, 88, 72, 91, and 77. To find the mean, or average, the scores are added and the sum is divided by 6 because there are 6 test scores:

$$\frac{93 + 84 + 88 + 72 + 91 + 77}{6} = \frac{505}{6} = 84.17$$

Given the mean of a data set and the sum of the data points, the sample size can be determined by dividing the sum by the mean. Suppose you are told that Kate averaged 12 points per game and scored a total of 156 points for the season. The number of games that she played (the sample size or the number of data points) can be determined by dividing the total points (sum of data points) by her average (mean of data points): $\frac{156}{12} = 13$. Therefore, Kate played in 13 games this season.

If given the mean of a data set and the sample size, the sum of the data points can be determined by multiplying the mean and sample size. Suppose you are told that Tom worked 6 days last week for an average of 5.5 hours per day. The total number of hours worked for the week (sum of data points) can be determined by multiplying his daily average (mean of data points) by the number of days worked (sample size): $5.5 \times 6 = 33$. Therefore, Tom worked a total of 33 hours last week.

The median of a data set is the value of the data point in the middle when the sample is arranged in numerical order. To find the median of a data set, the values are written in order from least to greatest. The lowest and highest values are simultaneously eliminated, repeating until the value in the middle remains. Suppose the salaries of math teachers are: $35,000; $38,500; $41,000; $42,000; $42,000; $44,500; $49,000. The values are listed from least to greatest to find the median. The lowest and highest values are eliminated until only the middle value remains. Repeating this step three times reveals a median salary of $42,000. If the sample set has an even number of data points, two values will remain after all others are eliminated. In this case, the mean of the two middle values is the median. Consider the following data set: 7, 9, 10, 13, 14, 14. Eliminating the lowest and highest values twice leaves two values, 10 and 13, in the middle. The mean of these values $\left(\frac{10+13}{2}\right)$ is the median. Therefore, the set has a median of 11.5.

The mode of a data set is the value that appears most often. A data set may have a single mode, multiple modes, or no mode. If different values repeat equally as often, multiple modes exist. If no value repeats, no mode exists. Consider the following data sets:

- A: 7, 9, 10, 13, 14, 14
- B: 37, 44, 33, 37, 49, 44, 51, 34, 37, 33, 44
- C: 173, 154, 151, 168, 155

Set A has a mode of 14. Set B has modes of 37 and 44. Set C has no mode.

The range of a data set is the difference between the highest and the lowest values in the set. The range can be considered the span of the data set. To determine the range, the smallest value in the set is subtracted from the largest value. The ranges for the data sets A, B, and C above are calculated as follows: A: $14 - 7 = 7$; B: $51 - 33 = 18$; C: $173 - 151 = 22$.

Best Description of a Set of Data

Measures of central tendency, namely mean, median, and mode, describe characteristics of a set of data. Specifically, they are intended to represent a *typical* value in the set by identifying a central position of the set. Depending on the characteristics of a specific set of data, different measures of central tendency are more indicative of a typical value in the set.

When a data set is grouped closely together with a relatively small range and the data is spread out somewhat evenly, the mean is an effective indicator of a typical value in the set. Consider the following data set representing the height of sixth grade boys in inches: 61 inches, 54 inches, 58 inches, 63 inches, 58 inches. The mean of the set is 58.8 inches. The data set is grouped closely (the range is only 9 inches) and the values are spread relatively evenly (three values below the mean and two values above the mean). Therefore, the mean value of 58.8 inches is an effective measure of central tendency in this case.

When a data set contains a small number of values either extremely large or extremely small when compared to the other values, the mean is not an effective measure of central tendency. Consider the following data set representing annual incomes of homeowners on a given street: $71,000; $74,000; $75,000; $77,000; $340,000. The mean of this set is $127,400. This figure does not indicate a typical value in the set, which contains four out of five values between $71,000 and $77,000. The median is a much more effective measure of central tendency for data sets such as these. Finding the middle value diminishes the influence of outliers, or numbers that may appear out of place, like the $340,000 annual income. The median for this set is $75,000 which is much more typical of a value in the set.

The mode of a data set is a useful measure of central tendency for categorical data when each piece of data is an option from a category. Consider a survey of 31 commuters asking how they get to work with results summarized below.

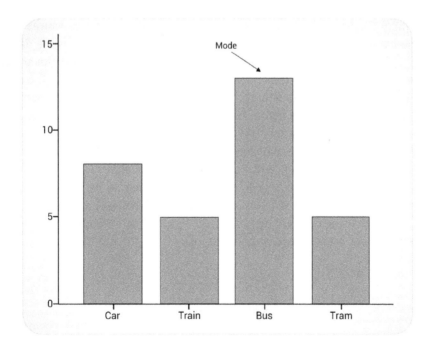

The mode for this set represents the value, or option, of the data that repeats most often. This indicates that the bus is the most popular method of transportation for the commuters.

Effects of Changes in Data

Changing all values of a data set in a consistent way produces predictable changes in the measures of the center and range of the set. A linear transformation changes the original value into the new value by either adding a given number to each value, multiplying each value by a given number, or both. Adding (or subtracting) a given value to each data point will increase (or decrease) the mean, median, and any modes by the same value. However, the range will remain the same due to the way that range is calculated. Multiplying (or dividing) a given value by each data point will increase (or decrease) the mean, median, and any modes, and the range by the same factor.

Consider the following data set, call it set *P*, representing the price of different cases of soda at a grocery store: $4.25, $4.40, $4.75, $4.95, $4.95, $5.15. The mean of set *P* is $4.74. The median is $4.85. The mode of the set is $4.95. The range is $0.90. Suppose the state passes a new tax of $0.25 on every case of soda sold. The new data set, set *T*, is calculated by adding $0.25 to each data point from set *P*. Therefore, set *T* consists of the following values: $4.50, $4.65, $5.00, $5.20, $5.20, $5.40. The mean of set *T* is $4.99. The median is $5.10. The mode of the set is $5.20. The range is $.90. The mean, median and mode of set *T* is equal to $0.25 added to the mean, median, and mode of set *P*. The range stays the same.

Now suppose, due to inflation, the store raises the cost of every item by 10 percent. Raising costs by 10 percent is calculated by multiplying each value by 1.1. The new data set, set *I*, is calculated by multiplying each data point from set *T* by 1.1. Therefore, set *I* consists of the following values: $4.95, $5.12, $5.50, $5.72, $5.72, $5.94. The mean of set *I* is $5.49. The median is $5.61. The mode of the set is $5.72. The range is $0.99. The mean, median, mode, and range of set *I* is equal to 1.1 multiplied by the mean, median, mode, and range of set *T* because each increased by a factor of 10 percent.

Comparing Data

Data sets can be compared by looking at the center and spread of each set. Measures of central tendency involve median, mean, midrange, and mode. The *mode* of a data set is the data value or values that appears the most frequently. The *midrange* is equal to the maximum value plus the minimum value divided by two. The *median* is the value that is halfway into each data set; it splits the data into two intervals. The *mean* is the sum of all data values divided by the number of data points. Two completely different sets of data can have the same mean. For example, a data set having values ranging from 0 to 100 and a data set having values ranging from 44 to 46 could both have means equal to 50. The first data set would have a much wider range, which is known as the *spread* of the data. It measures how varied the data is within each set. Spread can be defined further as either interquartile range or standard deviation. The *interquartile range (IQR)* is the range of the middle fifty percent of the data set. The *standard deviation, s,* quantifies the amount of variation with respect to the mean. A lower standard deviation shows that the data set does not differ much from the mean. A larger standard deviation shows that the data set is spread out farther away from the mean. The formula for standard deviation is:

$$ s = \sqrt{\frac{\sum(x - \bar{x})^2}{n - 1}} $$

where *x* is each value in the data set, $\bar{x}$ is the mean, and *n* is the total number of data points in the sample set. The square of the standard deviation is known as the *variance* of the data set. A data set can have outliers, and measures of central tendency that are not affected by outliers are the mode and median. Those measures are labeled as resistant measures of center.

Summarize Categorical Data for Two Categories in Two-Way Frequency Tables and Interpret Relative Frequencies

Two-Way Frequency Tables

Data that isn't described using numbers is known as *categorical data*. For example, age is numerical data but hair color is categorical data. Categorical data is summarized using two-way frequency tables. A *two-way frequency table* counts the relationship between two sets of categorical data. There are rows and columns for each category, and each cell represents frequency information that shows the actual data count between each combination.

For example, the graphic on the left-side below is a two-way frequency table showing the number of girls and boys taking language classes in school. Entries in the middle of the table are known as the *joint frequencies*. For example, the number of girls taking French class is 12, which is a joint frequency. The totals are the *marginal frequencies*. For example, the total number of boys is 20, which is a marginal frequency. If the frequencies are changed into percentages based on totals, the table is known as a *two-way relative frequency table*. Percentages can be calculated using the table total, the row totals, or the column totals. Here's the process of obtaining the two-way relative frequency table using the table total:

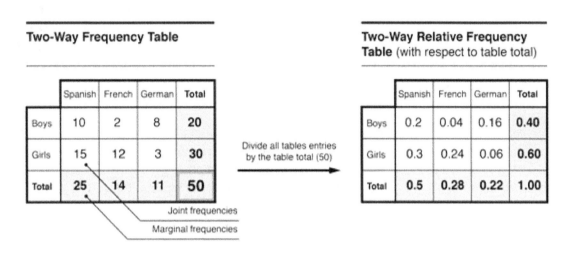

The middle entries are known as *joint probabilities* and the totals are *marginal probabilities*. In this data set, it appears that more girls than boys take Spanish class. However, that might not be the case because more girls than boys were surveyed and the results might be misleading. To avoid such errors, *conditional relative frequencies* are used. The relative frequencies are calculated based on a row or column.

Here are the conditional relative frequencies using column totals:

Two-Way Frequency Table

	Spanish	French	German	Total
Boys	10	2	8	20
Girls	15	12	3	30
Total	25	14	11	50

Divide each column entry by that column's total →

Two-Way Relative Frequency Table (with respect to table total)

	Spanish	French	German	Total
Boys	0.4	0.14	0.73	0.4
Girls	0.6	0.86	0.27	0.6
Total	1.00	1.00	1.00	1.00

Data Conclusions

Two-way frequency tables can help in making many conclusions about the data. If either the row or column of conditional relative frequencies differs between each row or column of the table, then an association exists between the two categories. For example, in the above tables, the majority of boys are taking German while the majority of girls are taking French. If the frequencies are equal across the rows, there is no association and the variables are labelled as independent. It's important to note that the association does exist in the above scenario, though these results may not occur the next semester when students are surveyed.

Plotting Variables

A *scatter plot* is a way to visually represent the relationship between two variables. Each variable has its own axis, and usually the independent variable is plotted on the horizontal axis while the dependent variable is plotted on the vertical axis. Data points are plotted in a process that's similar to how ordered pairs are plotted on an *xy*-plane. Once all points from the data set are plotted, the scatter plot is finished. Below is an example of a scatter plot that's plotting the quality and price of an item. Note that price is the independent variable and quality is the dependent variable:

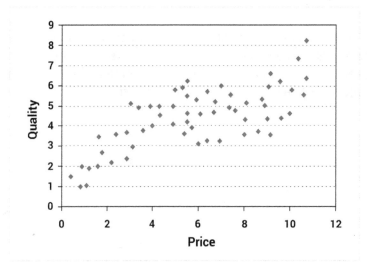

In this example, the quality of the item increases as the price increases.

Interpret the Slope and the Intercept of a Linear Model in the Context of the Data

Linear Regression

Regression lines are a way to calculate a relationship between the independent variable and the dependent variable. A straight line means that there's a linear trend in the data. Technology can be used to find the equation of this line (e.g., a graphing calculator or Microsoft Excel®). In either case, all of the data points are entered, and a line is "fit" that best represents the shape of the data. Other functions used to model data sets include quadratic and exponential models.

Estimating Data Points

Regression lines can be used to estimate data points not already given. For example, if an equation of a line is found that fit the temperature and beach visitor data set, its input is the average daily temperature and its output is the projected number of visitors. Thus, the number of beach visitors on a 100-degree day can be estimated. The output is a data point on the regression line, and the number of daily visitors is expected to be greater than on a 96-degree day because the regression line has a positive slope.

Plotting and Analyzing Residuals

Once the function is found that fits the data, its accuracy can be calculated. Therefore, how well the line fits the data can be determined. The difference between the actual dependent variable from the data set and the estimated value located on the regression line is known as a *residual*. Therefore, the residual is known as the predicted value $\hat{y}$ minus the actual value y. A residual is calculated for each data point and can be plotted on the scatterplot. If all the residuals appear to be approximately the same distance from the regression line, the line is a good fit. If the residuals seem to differ greatly across the board, the line isn't a good fit.

Interpreting the Regression Line

The formula for a regression line is $y = mx + b$, where m is the slope and b is the y-intercept. Both the slope and y-intercept are found in the *Method of Least Squares*, which is the process of finding the equation of the line through minimizing residuals. The slope represents the rate of change in y as x gets larger. Therefore, because y is the dependent variable, the slope actually provides the predicted values given the independent variable. The y-intercept is the predicted value for when the independent variable equals zero. In the temperature example, the y-intercept is the expected number of beach visitors for a very cold average daily temperature of zero degrees.

Correlation Coefficient

The *correlation coefficient (r)* measures the association between two variables. Its value is between -1 and 1, where -1 represents a perfect negative linear relationship, 0 represents no relationship, and 1 represents a perfect positive linear relationship. A *negative linear relationship* means that as x values increase, y values decrease. A *positive linear relationship* means that as x values increase, y values increase. The formula for computing the correlation coefficient is:

$$r = \frac{n(\sum xy) - (\sum x)(\sum y)}{\sqrt{n(\sum x^2) - (\sum x)^2}\sqrt{n(\sum y^2) - (\Sigma y)^2}}$$

n is the number of data points

Both Microsoft Excel® and a graphing calculator can evaluate this easily once the data points are entered. A correlation greater than 0.8 or less than -0.8 is classified as "strong" while a correlation between -0.5 and 0.5 is classified as "weak."

Data Gathering Techniques

The three most common types of data gathering techniques are sample surveys, experiments, and observational studies. *Sample surveys* involve collecting data from a random sample of people from a desired population. The measurement of the variable is only performed on this set of people. To have accurate data, the sampling must be unbiased and random. For example, surveying students in an advanced calculus class on how much they enjoy math classes is not a useful sample if the population should be all college students based on the research question. An *experiment* is the method in which a hypothesis is tested using a trial-and-error process. A cause and the effect of that cause are measured, and the hypothesis is accepted or rejected. Experiments are usually completed in a controlled environment where the results of a control population are compared to the results of a test population. The groups are selected using a randomization process in which each group has a representative mix of the population being tested. Finally, an *observational study* is similar to an experiment. However, this design is used when there cannot be a designed control and test population because of circumstances (e.g., lack of funding or unrealistic expectations). Instead, existing control and test populations must be used, so this method has a lack of randomization.

Population Mean and Proportion

Both the population mean and proportion can be calculated using data from a sample. The *population mean (μ)* is the average value of the parameter for the entire population. Due to size constraints, finding the exact value of μ is impossible, so the mean of the sample population is used as an estimate instead. The larger the sample size, the closer the sample mean gets to the population mean. An alternative to finding μ is to find the *proportion* of the population, which is the part of the population with the given characteristic. The proportion can be expressed as a decimal, a fraction, or a percentage, and can be given as a single value or a range of values. Because the population mean and proportion are both estimates, there's a *margin of error*, which is the difference between the actual value and the expected value.

T-Tests

A *randomized experiment* is used to compare two treatments by using statistics involving a *t-test*, which tests whether two data sets are significantly different from one another. To use a t-test, the test statistic must follow a normal distribution. The first step of the test involves calculating the *t* value, which is given as:

$$t = \frac{\overline{x_1} - \overline{x_2}}{s_{\bar{x}_1 - \bar{x}_2}}$$

where $\bar{x}_1$ and $\bar{x}_2$ are the averages of the two samples. Also,

$$s_{\bar{x}_1 - \bar{x}_2} = \sqrt{\frac{s_1^2}{n_1} + \frac{s_2^2}{n_2}}$$

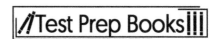

where s_1 and s_2 are the standard deviations of each sample and n_1 and n_2 are their respective sample sizes. The *degrees of freedom* for two samples are calculated as the following (rounded to the lowest whole number).

$$df = \frac{(n_1 - 1) + (n_2 - 1)}{2}$$

Also, a significance level α must be chosen, where a typical value is $\alpha = 0.05$. Once everything is compiled, the decision is made to use either a *one-tailed test* or a *two-tailed test*. If there's an assumed difference between the two treatments, a one-tailed test is used. If no difference is assumed, a two-tailed test is used.

Analyzing Test Results

Once the type of test is determined, the t-value, significance level, and degrees of freedom are applied to the published table showing the t distribution. The row is associated with degrees of freedom and each column corresponds to the probability. The t-value can be exactly equal to one entry or lie between two entries in a row. For example, consider a t-value of 1.7 with degrees of freedom equal to 30. This *test statistic* falls between the p values of 0.05 and 0.025. For a one-tailed test, the corresponding p value lies between 0.05 and 0.025. For a two-tailed test, the p values need to be doubled so the corresponding p value falls between 0.1 and 0.05. Once the probability is known, this range is compared to α. If $p < \alpha$, the hypothesis is rejected. If $p > \alpha$, the hypothesis isn't rejected. In a two-tailed test, this scenario means the hypothesis is accepted that there's no difference in the two treatments. In a one-tailed test, the hypothesis is accepted, indicating that there's a difference in the two treatments.

Evaluating Completed Tests

In addition to applying statistical techniques to actual testing, evaluating completed tests is another important aspect of statistics. Reports can be read that already have conclusions, and the process can be evaluated using learned concepts. For example, deciding if a sample being used is appropriate. Other things that can be evaluated include determining if the samples are randomized or the results are significant. Once statistical concepts are understood, the knowledge can be applied to many applications.

Distinguish Between Correlation and Causation

In an experiment, variables are the key to analyzing data, especially when data is in a graph or table. Variables can represent anything, including objects, conditions, events, and amounts of time.

Covariance is a general term referring to how two variables move in relation to each other. Take for example an employee that gets paid by the hour. For them, hours worked and total pay have a positive covariance. As hours worked increases, so does pay.

Constant variables remain unchanged by the scientist across all trials. Because they are held constant for all groups in an experiment, they aren't being measured in the experiment, and they are usually ignored. Constants can either be controlled by the scientist directly like the nutrition, water, and sunlight given to plants, or they can be selected by the scientist specifically for an experiment like using a certain animal species or choosing to investigate only people of a certain age group.

Independent variables are also controlled by the scientist, but they are the same only for each group or trial in the experiment. Each group might be composed of students that all have the same color of car or each trial may be run on different soda brands. The independent variable of an experiment is what is being indirectly tested because it causes change in the dependent variables.

Dependent variables experience change caused by the independent variable and are what is being measured or observed. For example, college acceptance rates could be a dependent variable of an experiment that sorted a large sample of high school students by an independent variable such as test scores. In this experiment, the scientist groups the high school students by the independent variable (test scores) to see how it affects the dependent variable (their college acceptance rates).

Note that most variables can be held constant in one experiment but independent or dependent in another. For example, when testing how well a fertilizer aids plant growth, its amount of sunlight should be held constant for each group of plants, but if the experiment is being done to determine the proper amount of sunlight a plant should have, the amount of sunlight is an independent variable because it is necessarily changed for each group of plants.

Correlation

An *X-Y diagram*, also known as a scatter diagram, visually displays the relationship between two variables. The independent variable is placed on the *x-axis*, or horizontal axis, and the dependent variable is placed on the *y-axis*, or vertical axis.

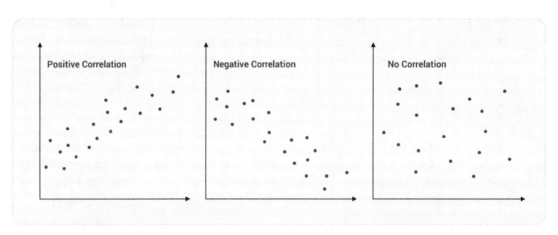

As shown in the figures above, an X-Y diagram may result in positive, negative, or no correlation between the two variables. So in the first scatter plot as the Y factor increases the X factor increases as well. The opposite is true as well: as the X factor increases the Y factor also increases. Thus, there is a positive correlation because one factor appears to positively affect the other factor

Correlation Versus Causation

It's important to note, however, that a positive correlation between two variables doesn't equate to a cause-and-effect relationship. For example, a positive correlation between labor hours and units produced may not equate to a cause and effect relationship between the two. Any instance of correlation only indicates how likely the presence of one variable is in the instance of another. The variables should be further analyzed to determine which, if any, other variables (i.e. quality of employee work) may contribute to the positive correlation.

Correlation and causation have two different meanings. If two values are correlated, there is an association between them. However, correlation doesn't necessarily mean that one variable causes the other. *Causation* (or "cause and effect") occurs when one variable causes the other. Average daily

temperature and number of beachgoers are correlated and have causation. If the temperature increases, the change in weather causes more people to go to the beach. However, alcoholism and smoking are correlated but don't have causation. The more someone drinks the more likely they are to smoke, but drinking alcohol doesn't cause someone to smoke.

Practice Questions

1. For the following similar triangles, what are the values of x and y (rounded to one decimal place)?

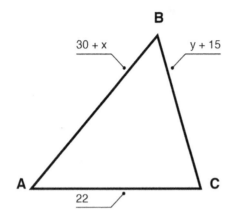 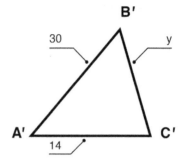

    a. $x = 16.5, y = 25.1$
    b. $x = 19.5, y = 24.1$
    c. $x = 17.1, y = 26.3$
    d. $x = 26.3, y = 17.1$

2. What are the center and radius of a circle with equation $4x^2 + 4y^2 - 16x - 24y + 51 = 0$?
    a. Center (3, 2) and radius ½
    b. Center (2, 3) and radius ½
    c. Center (3, 2) and radius ¼
    d. Center (2, 3) and radius ¼

3. If the ordered pair $(-3, -4)$ is reflected over the x-axis, what's the new ordered pair?
    a. $(-3, -4)$
    b. $(3, -4)$
    c. $(3, 4)$
    d. $(-3, 4)$

4. If the volume of a sphere is 288π cubic meters, what are the radius and surface area of the same sphere?
    a. Radius 6 meters and surface area 144π square meters
    b. Radius 36 meters and surface area 144π square meters
    c. Radius 6 meters and surface area 12π square meters
    d. Radius 36 meters and surface area 12π square meters

5. The triangle shown below is a right triangle. What's the value of x?

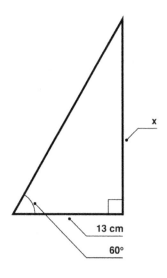

13 cm

60°

    a. $x = 1.73$
    b. $x = 0.57$
    c. $x = 13$
    d. $x = 22.49$

6. What's the midpoint of a line segment with endpoints $(-1, 2)$ and $(3, -6)$?
    a. $(1, 2)$
    b. $(1, 0)$
    c. $(-1, 2)$
    d. $(1, -2)$

7. What is the perimeter of the figure below? Note that the solid outer line is the perimeter.

8 in.

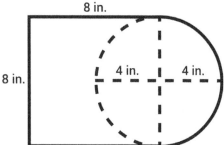

8 in.   4 in.   4 in.

    a. 48.565 in
    b. 36.565 in
    c. 39.78 in
    d. 39.565 in

8. Given the following triangle, what's the length of the missing side? Round the answer to the nearest tenth.

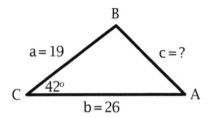

a. 17.0
b. 17.4
c. 18.0
d. 18.4

9. A rectangle has a length that is 5 feet longer than three times its width. If the perimeter is 90 feet, what is the length in feet?

a. 10
b. 20
c. 25
d. 35

10. An equilateral triangle has a perimeter of 18 feet. If a square whose sides have the same length as one side of the triangle is built, what will be the area of the square?

a. 6 square feet
b. 36 square feet
c. 256 square feet
d. 1000 square feet

11. What is the volume of a cone, in terms of π, with a radius of 10 centimeters and height of 12 centimeters?

a. 400 cm³
b. 200 cm³
c. 120 cm³
d. 140 cm³

12. What is the length of the other leg of a right triangle with a hypotenuse of 10 inches and a leg of 8 inches?

a. 6 in
b. 18 in
c. 80 in
d. 13 in

13. Using trigonometric ratios for a right angle, what is the value of the closest angle whose adjacent side is equal to 7.071 centimeters and whose hypotenuse is equal to 10 centimeters?

a. 15°
b. 30°
c. 45°
d. 90°

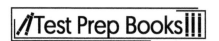

14. Using trigonometric ratios, what is the value of the other angle whose opposite side is equal to 1 inch and whose adjacent side is equal to the square root of 3 inches?

    a. 15°

    b. 30°

    c. 45°

    d. 90°

15. A farmer owns two (non-adjacent) square plots of land, which he wishes to fence. The area of one is 1000 square feet, while the area of the other is 10 square feet. How much fencing does he need, in feet?

    a. 44

    b. $40\sqrt{10}$

    c. $440\sqrt{10}$

    d. $44\sqrt{10}$

16. The graph shows the position of a car over a 10-second time interval. Which of the following is the correct interpretation of the graph for the interval 1 to 3 seconds?

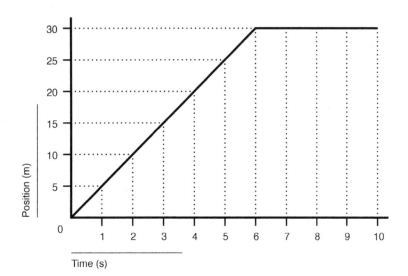

Time (s)

    a. The car remains in the same position.

    b. The car is traveling at a speed of 5m/s.

    c. The car is traveling up a hill.

    d. The car is traveling at 5mph.

17. A student gets an 85% on a test with 20 questions. How many answers did the student solve correctly?

    a. 15

    b. 16

    c. 17

    d. 18

18. Four people split a bill. The first person pays for $\frac{1}{5}$, the second person pays for $\frac{1}{4}$, and the third person pays for $\frac{1}{3}$. What fraction of the bill does the fourth person pay?

    a. $\frac{13}{60}$

    b. $\frac{47}{60}$

    c. $\frac{1}{4}$

    d. $\frac{4}{15}$

19. 6 is 30% of what number?

    a. 18
    b. 20
    c. 24
    d. 26

20. A closet is filled with red, blue, and green shirts. If $\frac{1}{3}$ of the shirts are green and $\frac{2}{5}$ are red, what fraction of the shirts are blue?

    a. $\frac{4}{15}$

    b. $\frac{1}{5}$

    c. $\frac{7}{15}$

    d. $\frac{1}{2}$

21. Shawna buys $2\frac{1}{2}$ gallons of paint. If she uses $\frac{1}{3}$ of it on the first day, how much does she have left?

    a. $1\frac{5}{6}$ gallons

    b. $1\frac{1}{2}$ gallons

    c. $1\frac{2}{3}$ gallons

    d. 2 gallons

22. You measure the width of your door to be 36 inches. The true width of the door is 35.75 inches. What is the relative error in your measurement?

    a. 0.7%
    b. 0.007%
    c. 0.99%
    d. 0.1%

23. What are all the factors of 12?
    a. 12, 24, 36
    b. 1, 2, 4, 6, 12
    c. 12, 24, 36, 48
    d. 1, 2, 3, 4, 6, 12

24. At the beginning of the day, Xavier has 20 apples. At lunch, he meets his sister Emma and gives her half of his apples. After lunch, he stops by his neighbor Jim's house and gives him 6 of his apples. He then uses ¾ of his remaining apples to make an apple pie for dessert at dinner. At the end of the day, how many apples does Xavier have left?
    a. 4
    b. 6
    c. 2
    d. 1

25. How many kilometers is 4382 feet?
    a. 1.336 kilometers
    b. 14,376 kilometers
    c. 1.437 kilometers
    d. 13,336 kilometers

26. Find the determinant of the matrix $\begin{bmatrix} -4 & 2 \\ 3 & -1 \end{bmatrix}$.
    a. -10
    b. -2
    c. 0
    d. 2

27. If $a \neq b$, solve for x if $\frac{1}{x} + \frac{2}{a} = \frac{2}{b}$
    a. $\frac{a-b}{ab}$
    b. $\frac{ab}{2(a-b)}$
    c. $\frac{2(a-b)}{ab}$
    d. $\frac{a-b}{2ab}$

28. If $x^2 + x - 3 = 0$, then $\left(x + \frac{1}{2}\right)^2 =$
    a. $\frac{11}{2}$
    b. $\frac{13}{4}$
    c. 11
    d. $\frac{121}{4}$

29. How do you solve $V = lwh$ for h?

    a. $lwV = h$

    b. $h = \dfrac{V}{lw}$

    c. $h = \dfrac{Vl}{w}$

    d. $h = \dfrac{Vw}{l}$

30. Which of the following equations best represents the problem below?
The width of a rectangle is 2 centimeters less than the length. If the perimeter of the rectangle is 44 centimeters, then what are the dimensions of the rectangle?

    a. $2l + 2(l - 2) = 44$
    b. $l + 2) + (l + 2) + l = 48$
    c. $l \times (l - 2) = 44$
    d. $(l + 2) + (l + 2) + l = 44$

31. Jessica buys 10 cans of paint. Red paint costs \$1 per can and blue paint costs \$2 per can. In total, she spends \$16. How many red cans did she buy?

    a. 2
    b. 3
    c. 4
    d. 5

32. Which of the following is the result of simplifying the expression:

$$\frac{4a^{-1}b^3}{a^4b^{-2}} \times \frac{3a}{b}$$

    a. $12a^3b^5$

    b. $12\dfrac{b^4}{a^4}$

    c. $\dfrac{12}{a^4}$

    d. $7\dfrac{b^4}{a}$

33. Simplify: $2x(3x + 1) - 5(3x + 1)$

    a. $10x(3x + 1)$

    b. $10x^2(3x + 1)$

    c. $(2x - 5)(3x + 1)$

    d. $(2x + 1)(3x - 5)$

34. If $4x - 3 = 5$, then $x =$
    a. 1
    b. 2
    c. 3
    d. 4

35. A root of $x^2 - 2x - 2$ is
    a. $1 + \sqrt{3}$
    b. $1 + 2\sqrt{2}$
    c. $2 + 2\sqrt{3}$
    d. $2 - 2\sqrt{3}$

36. A line goes through the point (-4, 0) and the point (0,2). What is the slope of the line?
    a. 2

    b. 4

    c. $\frac{3}{2}$

    d. $\frac{1}{2}$

37. What are the zeros of the function: $f(x) = x^3 + 4x^2 + 4x$?
    a. -2
    b. 0, -2
    c. 2
    d. 0, 2

38. What is the solution to the following system of equations?
$$x^2 - 2x + y = 8$$
$$x - y = -2$$
    a. $(-2, 3)$
    b. There is no solution.
    c. $(-2, 0)\ (1, 3)$
    d. $(-2, 0)\ (3, 5)$

39. Which of the following shows the correct result of simplifying the following expression:
$$(7n + 3n^3 + 3) + (8n + 5n^3 + 2n^4)$$
    a. $9n^4 + 15n - 2$
    b. $2n^4 + 5n^3 + 15n - 2$
    c. $9n^4 + 8n^3 + 15n$
    d. $2n^4 + 8n^3 + 15n + 3$

40. What is the simplified quotient of the following expression?

$$\frac{5x^3}{3x^2y} \div \frac{25}{3y^9}$$

a. $\frac{125x}{9y^{10}}$

b. $\frac{x}{5y^8}$

c. $\frac{5}{xy^8}$

d. $\frac{xy^8}{5}$

41. If $x > 3$, then $\frac{x^2-6x+9}{x^2-x-6} =$

a. $\frac{x+2}{x-3}$

b. $\frac{x-2}{x-3}$

c. $\frac{x-3}{x+3}$

d. $\frac{x-3}{x+2}$

42. Let $f(x) = 2x + 1, g(x) = \frac{x-1}{4}$. Find $g(f(x))$.

a. $\frac{x+1}{2}$

b. $\frac{x}{2}$

c. $\frac{2x^2-x-1}{4}$

d. $3x$

43. If $f(x) = (\frac{1}{2})^x$ and $a < b$, then which of the following must be true?

a. $f(a) < f(b)$

b. $f(a) > f(b)$

c. $f(a) + f(b) = 0$

d. $3f(a) = f(b)$

44. What type of function is modeled by the values in the following table?

X	f(x)
1	2
2	4
3	8
4	16
5	32

a. Linear
b. Exponential
c. Quadratic
d. Cubic

45. A line that travels from the bottom-left of a graph to the upper-right of the graph indicates what kind of relationship between a predictor and a dependent variable?
a. Positive
b. Negative
c. Exponential
d. Logarithmic

46. Which of the following is the best description of the relationship between Y and X?
a. The data has normal distribution.
b. X and Y have a negative relationship.
c. No relationship
d. X and Y have a positive relationship.

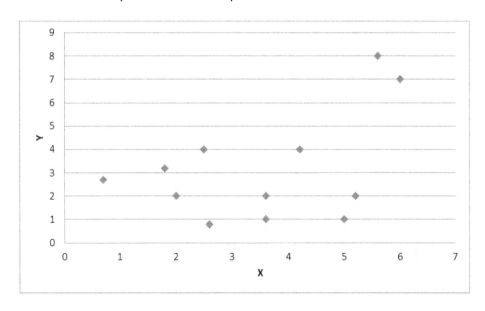

47. What is the slope of this line?

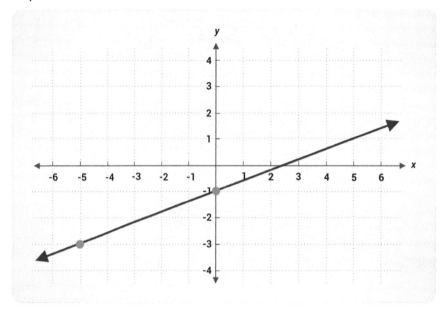

a. 2

b. $\frac{5}{2}$

c. $\frac{1}{2}$

d. $\frac{2}{5}$

48. Which graph will be a line parallel to the graph of $y = 3x - 2$?
    a. $2y - 6x = 2$
    b. $y - 4x = 4$
    c. $3y = x - 2$
    d. $2x - 2y = 2$

49. An equation for the line passing through the origin and the point $(2, 1)$ is
    a. $y = 2x$
    b. $y = \frac{1}{2}x$
    c. $y = x - 2$
    d. $2y = x + 1$

50. For which real numbers x is $-3x^2 + x - 8 > 0$?
    a. All real numbers x
    b. $-2\sqrt{\frac{2}{3}} < x < 2\sqrt{\frac{2}{3}}$
    c. $1 - 2\sqrt{\frac{2}{3}} < x < 1 + 2\sqrt{\frac{2}{3}}$
    d. For no real numbers x

51. What is the domain for the function $y = \sqrt{x}$?
    a. All real numbers
    b. $x \geq 0$
    c. $x > 0$
    d. $y \geq 0$

52. The phone bill is calculated each month using the equation $c = 50g + 75$. The cost of the phone bill per month is represented by c, and g represents the gigabytes of data used that month. What is the value and interpretation of the slope of this equation?
    a. 75 dollars per day
    b. 75 gigabytes per day
    c. 50 dollars per day
    d. 50 dollars per gigabyte

53. What is the inverse of the function $f(x) = 3x - 5$?
    a. $f^{-1}(x) = \frac{x}{3} + 5$
    b. $f^{-1}(x) = \frac{5x}{3}$
    c. $f^{-1}(x) = 3x + 5$
    d. $f^{-1}(x) = \frac{x+5}{3}$

54. What are the zeros of $f(x) = x^2 + 4$?
    a. $x = -4$
    b. $x = \pm 2i$
    c. $x = \pm 2$
    d. $x = \pm 4i$

55. What are the y-intercept(s) for $y = x^2 + 3x - 4$?
    a. $y = 1$
    b. $y = -4$
    c. $y = 3$
    d. $y = 4$

56. Is the following function even, odd, neither, or both?
$$y = \frac{1}{2}x^4 + 2x^2 - 6$$
    a. Even
    b. Odd
    c. Neither
    d. Both

57. Which equation is not a function?
    a. $y = |x|$
    b. $y = \sqrt{x}$
    c. $x = 3$
    d. $y = 4$

58. How could the following function be rewritten to identify the zeros?

$$y = 3x^3 + 3x^2 - 18x$$

a. $y = 3x(x + 3)(x - 2)$
b. $y = x(x - 2)(x + 3)$
c. $y = 3x(x - 3)(x + 2)$
d. $y = (x + 3)(x - 2)$

59. What is the slope of the line tangent to the graph of $y = x^3 - 4$ at the point where $x = 2$?

a. $3x^2$
b. 4
c. -4
d. 12

60. For a group of 20 men, the median weight is 180 pounds and the range is 30 pounds. If each man gains 10 pounds, which of the following would be true?

a. The median weight will increase, and the range will remain the same.
b. The median weight and range will both remain the same.
c. The median weight will stay the same, and the range will increase.
d. The median weight and range will both increase.

61. A sample data set contains the following values: 1, 3, 5, 7. What's the standard deviation of the set?

a. 2.58
b. 4
c. 6.23
d. 1.1

62. In a statistical experiment, 29 college students are given an exam during week 11 of the semester, and 30 college students are given an exam during week 12 of the semester. Both groups are being tested to determine which exam week might result in a higher grade. What's the degree of freedom in this experiment?

a. 29
b. 30
c. 59
d. 28

63. Dwayne has received the following scores on his math tests: 78, 92, 83, 97. What score must Dwayne get on his next math test to have an overall average of 90?

a. 89
b. 98
c. 95
d. 100

64. What is the overall median of Dwayne's current scores: 78, 92, 83, 97?

a. 19
b. 85
c. 83
d. 87.5

65. Which measure for the center of a small sample set would be most affected by outliers?
    a. Mean
    b. Median
    c. Mode
    d. Range

66. Given the value of a given stock at monthly intervals, which graph should be used to best represent the trend of the stock?
    a. Box plot
    b. Line plot
    c. Line graph
    d. Circle graph

67. A sample data set contains the following values: 1, 3, 5, 7. What's the standard deviation of the set?
    a. 2.58
    b. 4
    c. 6.23
    d. 1.1

68. The following graph compares the various test scores of the top three students in each of these teacher's classes. Based on the graph, which teacher's students had the lowest range of test scores?

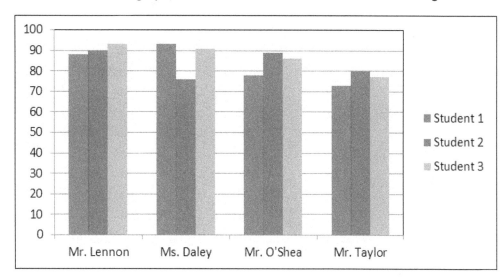

    a. Mr. Lennon
    b. Mr. O'Shea
    c. Mr. Taylor
    d. Ms. Daley

Answer Explanations

1. C: Because the triangles are similar, the lengths of the corresponding sides are proportional. Therefore:

$$\frac{30 + x}{30} = \frac{22}{14} = \frac{y + 15}{y}$$

This results in the equation $14(30 + x) = 22 \cdot 30$ which, when solved, gives $x = 17.1$. The proportion also results in the equation $14(y + 15) = 22y$ which, when solved, gives $y = 26.3$.

2. B: The technique of completing the square must be used to change the original equation into the standard equation of a circle.

Start with the original equation:

$$4x^2 + 4y^2 - 16x - 24y + 51 = 0$$

First, the constant must be moved to the right-hand side of the equals sign, and each term must be divided by the coefficient of the x^2 term (which is 4). The x and y terms must be grouped together to obtain:

$$x^2 - 4x + y^2 - 6y = -\frac{51}{4}$$

Then, the process of completing the square must be completed for each variable. This gives:

$$(x^2 - 4x + 4) + (y^2 - 6y + 9)$$

$$-\frac{51}{4} + 4 + 9$$

The equation can be written as:

$$(x - 2)^2 + (y - 3)^2 = \frac{1}{4}$$

Therefore, the center of the circle is (2, 3) and the radius is $\sqrt{1/4} = 1/2$

3. D: When an ordered pair is reflected over an axis, the sign of one of the coordinates must change. When it's reflected over the x-axis, the sign of the x coordinate must change. The y value remains the same. Therefore, the new ordered pair is $(-3, 4)$.

4. A: Because the volume of the given sphere is 288π cubic meters, this means:

$$\frac{4}{3}\pi r^3 = 288\pi$$

This equation is solved for r to obtain a radius of 6 meters. The formula for the surface area of a sphere is $4\pi r^2$, so if $r = 6$ in this formula, the surface area is 144π square meters.

5. D: SOHCAHTOA is used to find the missing side length. Because the angle and adjacent side are known, $\tan 60 = \frac{x}{13}$. Making sure to evaluate tangent with an argument in degrees, this equation gives:

$$x = 13 \tan 60 = 13 \cdot 1.73 = 22.49$$

6. D: The midpoint formula should be used.

$$M = \left(\frac{x_1 + x_2}{2}, \frac{y_1 + y_2}{2}\right)$$

$$\left(\frac{-1 + 3}{2}, \frac{2 + (-6)}{2}\right) = (1, -2)$$

7. B: The figure is composed of three sides of a square and a semicircle. The sides of the square are simply added: 8 + 8 + 8 = 24 inches. The circumference of a circle is found by the equation C = 2πr. The radius is 4 in, so the circumference of the circle is 25.13 in. Only half of the circle makes up the outer border of the figure (part of the perimeter) so half of 25.13 in is 12.565 in. Therefore, the total perimeter is:

$$24 \text{ in} + 12.565 \text{ in} = 36.565 \text{ in}$$

The other answer choices use the incorrect formula or fail to include all of the necessary sides.

8. B: Because this isn't a right triangle, SOHCAHTOA can't be used. However, the law of cosines can be used. Therefore:

$$c^2 = a^2 + b^2 - 2ab$$

$$\cos C = 19^2 + 26^2 - 2 \cdot 19 \cdot 26 \cdot \cos 42°$$

$$302.773$$

Taking the square root and rounding to the nearest tenth results in $c = 17.4$.

9. D: Denote the width as w and the length as l. Then:

$$l = 3w + 5$$

The perimeter is:

$$2w + 2l = 90$$

Substituting the first expression for l into the second equation yields $2(3w + 5) + 2w = 90$, or $8w = 80$, so $l = 10$. Putting this into the first equation, it yields:

$$l = 3(10) + 5 = 35$$

10. B: An equilateral triangle has three sides of equal length, so if the total perimeter is 18 feet, each side must be 6 feet long. A square with sides of 6 feet will have an area of $6^2 = 36$ square feet.

Test Prep Books!!!

11. A: The volume of a cone is $(\pi r^2 h)$, divided by 3, and $(\pi \times 10^2 \times 12)$, divided by 3 is 400 cm³. Choice *B* is $10^2 \times 2$. Choice *C* is incorrect because it is 10×12. Choice *D* is also incorrect because that is:

$$10^2 + 40$$

12. A: This answer is correct because $100 - 64$ is 36, and taking the square root of 36 is 6. Choice *B* is not the correct answer because that is $10 + 8$. Choice *C* is not the correct answer because that is 8×10. Choice *D* is also not the correct answer because there is no reason to arrive at that number.

13. C: The cosine of 45° is equal to 0.7071. Choice *A* is not the correct answer because the cosine of 15° is .9659. Choice *B* is not the correct answer because the cosine of 30° is .8660. Choice *D* is not correct because the cosine of 90° is 0.

14. B: The tangent of 30° is 1 over the square root of 3. Choice *A* is not the correct answer because the tangent of 15° is 0.2679. Choice *C* is not the correct answer because the tangent of 45° is 1. Choice *D* is not the correct answer because the tangent of 90° is undefined.

15. D: The first field has an area of 1000 feet, so the length of one side is:

$$\sqrt{1000} = 10\sqrt{10}$$

Since there are four sides to a square, the total perimeter is $40\sqrt{10}$. The second square has an area of 10 square feet, so the length of one side is $\sqrt{10}$, and the total perimeter is $4\sqrt{10}$. Adding these together gives:

$$40\sqrt{10} + 4\sqrt{10} = (40 + 4)\sqrt{10}$$

$$44\sqrt{10}$$

16. B: The car is traveling at a speed of five meters per second. On the interval from one to three seconds, the position changes by fifteen meters. By making this change in position over time into a rate, the speed becomes ten meters in two seconds or five meters in one second.

17. C: 85% of a number means multiplying that number by 0.85. So, $0.85 \times 20 = \frac{85}{100} \times \frac{20}{1}$, which can be simplified to:

$$\frac{17}{20} \times \frac{20}{1} = 17$$

18. A: To find the fraction of the bill that the first three people pay, the fractions need to be added, which means finding common denominator. The common denominator will be 60.

$$\frac{1}{5} + \frac{1}{4} + \frac{1}{3} = \frac{12}{60} + \frac{15}{60} + \frac{20}{60} = \frac{47}{60}$$

The remainder of the bill is:

$$1 - \frac{47}{60} = \frac{60}{60} - \frac{47}{60} = \frac{13}{60}$$

19. B: 30% is 3/10. The number itself must be 10/3 of 6, or:

$$\frac{10}{3} \times 6 = 10 \times 2 = 20$$

20. A: The total fraction taken up by green and red shirts will be:

$$\frac{1}{3} + \frac{2}{5} = \frac{5}{15} + \frac{6}{15} = \frac{11}{15}$$

The remaining fraction is:

$$1 - \frac{11}{15} = \frac{15}{15} - \frac{11}{15} = \frac{4}{15}$$

21. C: If she has used 1/3 of the paint, she has 2/3 remaining. $2\frac{1}{2}$ gallons are the same as $\frac{5}{2}$ gallons. The calculation is:

$$\frac{2}{3} \times \frac{5}{2} = \frac{5}{3} = 1\frac{2}{3} \text{ gallons}$$

22. A: The relative error can be found by finding the absolute error and making it a percent of the true value. The absolute value is:

$$36 - 35.75 = 0.25$$

This error is then divided by 36—the true value—to find 0.7%.

23. D: 1, 2, 3, 4, 6, 12. A given number divides evenly by each of its factors to produce an integer (no decimals). The number 5, 7, 8, 9, 10, 11 (and their opposites) do not divide evenly into 12. Therefore, these numbers are not factors.

24. D: This problem can be solved using basic arithmetic. Xavier starts with 20 apples, then gives his sister half, so 20 divided by 2.

$$\frac{20}{2} = 10$$

He then gives his neighbor 6, so 6 is subtracted from 10.

$$10 - 6 = 4$$

Lastly, he uses ¾ of his apples to make an apple pie, so to find remaining apples, the first step is to subtract ¾ from one and then multiply the difference by 4.

$$\left(1 - \frac{3}{4}\right) \times 4 = ?$$

$$\left(\frac{4}{4} - \frac{3}{4}\right) \times 4 = ?$$

$$\left(\frac{1}{4}\right) \times 4 = 1$$

25. A: The conversion can be obtained by setting up and solving the following equation:

$$4382 \, ft \; \times \frac{0.3048 \, m}{1 \, ft} \times \frac{1 \, km}{1000 \, m} = 1.336 \, km$$

26. B: The determinant of a 2 x 2 matrix is $ad - bc$. The calculation is:

$$-4(-1) - 2(3) = 4 - 6 = -2$$

27. B: $\frac{2}{a}$ must be subtracted from both sides, with a result of:

$$\frac{1}{x} = \frac{2}{b} - \frac{2}{a}$$

The reciprocal of both sides needs to be taken, but the right-hand side needs to be written as a single fraction in order to do that. Since the two fractions on the right have denominators that are not equal, a common denominator of ab is needed. This leaves:

$$\frac{1}{x} = \frac{2a}{ab} - \frac{2b}{ab} = \frac{2(a - b)}{ab}$$

Taking the reciprocals, which can be done since $b - a$ is not zero, with a result of:

$$x = \frac{ab}{2(a - b)}$$

28. B: The first step is to use the quadratic formula on the first equation ($x^2 + x - 3 = 0$) to solve for x. In this case, a is 1, b is 1, and c is -3, yielding:

$$x = \frac{-b \pm \sqrt{b^2 - 4ac}}{2a}$$

$$x = \frac{-1 \pm \sqrt{1 - 4 \times 1(-3)}}{2}$$

$$x = \frac{-1}{2} \pm \frac{\sqrt{13}}{2}$$

Therefore, $x + \frac{1}{2}$, which is in our second equation, equals $\pm \frac{\sqrt{13}}{2}$. We are looking for $\left(x + \frac{1}{2}\right)^2$ though, so we square the $\pm \frac{\sqrt{13}}{2}$. Doing so causes the $\pm$ cancels and left with $\left(\frac{\sqrt{13}}{2}\right)^2 = \frac{13}{4}$

29. B: The formula can be manipulated by dividing both sides by the length, l, and the width, w. The length and width will cancel on the right, leaving height by itself.

30. A: The first step is to determine the unknown, which is in terms of the length, l.

The second step is to translate the problem into the equation using the perimeter of a rectangle:

$$P = 2l + 2w$$

The width is the length minus 2 centimeters. The resulting equation is:

$$2l + 2(l - 2) = 44$$

The equation can be solved as follows:

$2l + 2l - 4 = 44$	Apply the distributive property on the left side of the equation
$4l - 4 = 44$	Combine like terms on the left side of the equation
$4l = 48$	Add 4 to both sides of the equation
$l = 12$	Divide both sides of the equation by 4

The length of the rectangle is 12 centimeters. The width is the length minus 2 centimeters, which is 10 centimeters. Checking the answers for length and width forms the following equation:

$$44 = 2(12) + 2(10)$$

The equation can be solved using the order of operations to form a true statement: $44 = 44$.

31. C: We are trying to find x, the number of red cans. The equation can be set up like this:

$$x + 2(10 - x) = 16$$

The left x is actually multiplied by $1, the price per red can. Since we know Jessica bought 10 total cans, $10 - x$ is the number blue cans that she bought. We multiply the number of blue cans by $2, the price per blue can.

That should all equal $16, the total amount of money that Jessica spent. Working that out gives us:

$$x + 20 - 2x = 16$$

$$20 - x = 16$$

$$x = 4$$

32. B: To simplify the given equation, the first step is to make all exponents positive by moving them to the opposite place in the fraction. This expression becomes:

$$\frac{4b^3 b^2}{a^1 a^4} * \frac{3a}{b}$$

Then the rules for exponents can be used to simplify. Multiplying the same bases means the exponents can be added. Dividing the same bases means the exponents are subtracted.

33. C: The $(3x + 1)$ can be factored to get:

$$(2x - 5)(3x + 1)$$

34. B: Add 3 to both sides to get $4x = 8$. Then divide both sides by 4 to get $x = 2$.

35. A: Check each value, but it is easiest to use the quadratic formula, which gives:

$$x = \frac{2 \pm \sqrt{(-2)^2 - 4(1)(-2)}}{2} = 1 \pm \frac{\sqrt{12}}{2}$$

$$1 \pm \frac{2\sqrt{3}}{2} = 1 \pm \sqrt{3}$$

The only one of these which appears as an answer choice is:

$$1 + \sqrt{3}$$

36. D: The slope is given by the change in *y* divided by the change in *x*. The change in *y* is 2-0 = 2, and the change in *x* is 0 − (-4) = 4. The slope is:

$$\frac{2}{4} = \frac{1}{2}$$

37. B: There are two zeros for the given function. They are $x = 0, -2$. The zeros can be found a number of ways, but this particular equation can be factored into:

$$f(x) = x(x^2 + 4x + 4)$$

$$x(x + 2)(x + 2)$$

By setting each factor equal to zero and solving for x, there are two solutions. On a graph, these zeros can be seen where the line crosses the x-axis.

38. D: This system of equations involves one quadratic function and one linear function, as seen from the degree of each equation. One way to solve this is through substitution.

Solving for y in the second equation yields:

$$y = x + 2$$

Plugging this equation in for the y of the quadratic equation yields:

$$x^2 - 2x + x + 2 = 8$$

Simplifying the equation, it becomes:

$$x^2 - x + 2 = 8$$

Setting this equal to zero and factoring, it becomes:

$$x^2 - x - 6 = 0 = (x - 3)(x + 2)$$

Solving these two factors for x gives the zeros:

$$x = 3, -2$$

To find the y-value for the point, each number can be plugged in to either original equation. Solving each one for y yields the points $(3, 5)$ and $(-2, 0)$.

39. D: The expression is simplified by collecting like terms. Terms with the same variable and exponent are like terms, and their coefficients can be added.

40. D: Dividing rational expressions follows the same rule as dividing fractions. The division is changed to multiplication, and the reciprocal is found in the second fraction. This turns the expression into:

$$\frac{5x^3}{3x^2} \times \frac{3y^9}{25}$$

Multiplying across and simplifying, the final expression is $\frac{xy^8}{5}$.

41. D: Factor the numerator into $x^2 - 6x + 9 = (x - 3)^2$, since:

$$-3 - 3 = -6, (-3)(-3) = 9$$

Factor the denominator into $x^2 - x - 6 = (x - 3)(x + 2)$, since $-3 + 2 = -1, (-3)(2) = -6$. This means the rational function can be rewritten as:

$$\frac{x^2 - 6x + 9}{x^2 - x - 6} = \frac{(x - 3)^2}{(x - 3)(x + 2)}$$

Using the restriction of x > 3, do not worry about any of these terms being 0, and cancel an $x - 3$ from the numerator and the denominator, leaving $\frac{x-3}{x+2}$.

42. B: Recall that to compose functions, replace the x in the expression for g with the expression for f everywhere there is x. So:

$$g\big(f(x)\big) = \frac{f(x) - 1}{4}$$

$$\frac{2x + 1 - 1}{4} = \frac{2x}{4} = \frac{x}{2}$$

43. B: Here, f is an exponential function whose base is less than 1. In this function, f is always decreasing. This means that when a is less than b:

$$f(a) > f(b)$$

44. B: The table shows values that are increasing exponentially. The differences between the inputs are the same, while the differences in the outputs are changing by a factor of 2. The values in the table can be modeled by the equation:

$$f(x) = 2^x$$

45. A: This vector indicates a positive relationship. A negative relationship would show points traveling from the top-left of the graph to the bottom-right. Exponential and logarithmic functions aren't linear (don't create a straight line), so these options could have been immediately eliminated.

46. C: There is no verifiable relationship between the two variables. While it may seem to have somewhat of a positive correlation because of the last two data points: (5.6,8) and (6,7), you must also take into account the two data points before those (5,1) and (5.2, 2) that have low Y values despite high X values. Data with a normal distribution (Choice A) has an arc to it. This data does not.

47. D: The slope is given by the change in *y* divided by the change in *x*. Specifically, it's:

$$slope = \frac{y_2 - y_1}{x_2 - x_1}$$

The first point is (-5, -3), and the second point is (0, -1). Work from left to right when identifying coordinates. Thus the point on the left is point 1 (-5,-3) and the point on the right is point 2 (0,-1).

Now we need to just plug those numbers into the equation:

$$slope = \frac{-1 - (-3)}{0 - (-5)}$$

It can be simplified to:

$$slope = \frac{-1 + 3}{0 + 5}$$

$$slope = \frac{2}{5}$$

48. A: Parallel lines have the same slope. The slope of *C* can be seen to be 1/3 by dividing both sides by 3. The others are in standard form $Ax + By = C$, for which the slope is given by $\frac{-A}{B}$. The slope of A is 3; the slope of B is 4. The slope of *D* is 1.

49. B: The slope will be given by:

$$\frac{1 - 0}{2 - 0} = \frac{1}{2}$$

The *y*-intercept will be 0, since it passes through the origin. Using slope-intercept form, the equation for this line is:

$$y = \frac{1}{2}x$$

50. D: Because the coefficient of x^2 is negative, this function has a graph that is a parabola that opens downward. Therefore, it will be greater than 0 between its real roots, if it has any. Checking the discriminant, the result is:

$$1^2 - 4(-3)(-8)$$

$$1 - 96 = -95$$

Since the discriminant is negative, this equation has no real solutions. Since this has no real roots, it must be always positive or always negative. Its graph opens downward, so it has at least some negative values. That means it is always negative. Thus, it is greater than zero for no real numbers.

51. B: The domain is all possible input values, or x-values. For this equation, the domain is every number greater than or equal to zero. There are no negative numbers in the domain because taking the square root of a negative number results in an imaginary number.

52. D: The slope from this equation is 50, and it is interpreted as the cost per gigabyte used. Since the g-value represents number of gigabytes and the equation is set equal to the cost in dollars, the slope relates these two values. For every gigabyte used on the phone, the bill goes up 50 dollars.

53. A: The inverse of a function is found by switching the x and y in the equation and solving for y. In the given equation, solving for y is done by adding 5 to both sides, then dividing both sides by 3. This answer can be checked on the graph by verifying the lines are reflected over $y = x$.

54. B: The zeros of this function can be found by using the quadratic formula:

$$x = \frac{-b \pm \sqrt{b^2 - 4ac}}{2a}$$

Identifying a, b, and c can be done from the equation as well because it is in standard form. The formula becomes:

$$x = \frac{0 \pm \sqrt{0^2 - 4(1)(4)}}{2(1)} = \frac{\sqrt{-16}}{2}$$

Since there is a negative underneath the radical, the answer is a complex number: x= +/- 2i (but using equation editor).

55. B: The y-intercept of an equation is found where the x-value is zero. Plugging zero into the equation for x, the first two terms cancel out, leaving -4.

56. A: The equation is *even* because:

$$f(-x) = f(x)$$

Plugging in a negative value will result in the same answer as when plugging in the positive of that same value.

The function:

$$f(-2) = \frac{1}{2}(-2)^4 + 2(-2)^2 - 6$$

$$8 + 8 - 6 = 10$$

yields the same value as:

$$f(2) = \frac{1}{2}(2)^4 + 2(2)^2 - 6$$

$$8 + 8 - 6 = 10$$

57. C: The equation $x = 3$ is not a function because it does not pass the vertical line test. This test is made from the definition of a function, where each x-value must be mapped to one and only one y-value. This equation is a vertical line, so the x-value of 3 is mapped with an infinite number of y-values.

58. A: The function can be factored to identify the zeros. First, the term $3x$ is factored out to the front because each term contains $3x$. Then, the quadratic is factored into $(x + 3)(x - 2)$.

59. D: Finding the slope of the line tangent to the given function involves taking the derivative twice. The first derivative gives the line tangent to the graph. The second derivative finds the slope of that line. The line tangent to the graph has an equation $y' = 3x^2$. The slope of this line at $x = 2$ is found by the second derivative, $y = 6x$, or $y = 6(2) = 12$.

60. A: If each man gains 10 pounds, every original data point will increase by 10 pounds. Therefore, the man with the original median will still have the median value, but that value will increase by 10. The smallest value and largest value will also increase by 10 and, therefore, the difference between the two won't change. The range does not change in value and, thus, remains the same.

61. A: First, the sample mean must be calculated.

$$\bar{x} = \frac{1}{4}(1 + 3 + 5 + 7) = 4$$

The standard deviation of the data set is:

$$\sigma = \sqrt{\frac{\sum(x - \bar{x})^2}{n - 1}}$$

and $n = 4$ represents the number of data points. Therefore:

$$\sigma = \sqrt{\frac{1}{3}[(1 - 4)^2 + (3 - 4)^2 + (5 - 4)^2 + (7 - 4)^2]}$$

$$\sqrt{\frac{1}{3}(9 + 1 + 1 + 9)} = 2.58$$

62. D: The degree of freedom for two samples is calculated as $df = \frac{(n_1-1)+(n_2-1)}{2}$ rounded to the lowest whole number. For this example, $df = \frac{(29-1)+(30-1)}{2} = \frac{28+29}{2} = 28.5$ which, rounded to the lowest whole number, is 28.

63. D: To find the average of a set of values, add the values together and then divide by the total number of values. In this case, include the unknown value of what Dwayne needs to score on his next test, in order to solve it.

$$\frac{78 + 92 + 83 + 97 + x}{5} = 90$$

Add the unknown value to the new average total, which is 5. Then multiply each side by 5 to simplify the equation, resulting in:

$$78 + 92 + 83 + 97 + x = 450$$

$$350 + x = 450$$

$$x = 100$$

Dwayne would need to get a perfect score of 100 in order to get an average of at least 90.

Test this answer by substituting back into the original formula.

$$\frac{78 + 92 + 83 + 97 + 100}{5} = 90$$

64. D: For an even number of total values, the *median* is calculated by finding the *mean* or average of the two middle values once all values have been arranged in ascending order from least to greatest. In this case, $(92 + 83) \div 2$ would equal the median 87.5, answer *D*.

65. A: Mean. An outlier is a data value that's either far above or below the majority of values in a sample set. The mean is the average of all values in the set. In a small sample, a very high or low number could greatly change the average. The median is the middle value when arranged from lowest to highest. Outliers would have no more of an effect on the median than any other value. Mode is the value that repeats most often in a set. Assuming that the same outlier doesn't repeat, outliers would have no effect on the mode of a sample set.

66. C: The scenario involves data consisting of two variables: month and stock value. Box plots display data consisting of values for one variable. Therefore, a box plot is not an appropriate choice. Both line plots and circle graphs are used to display frequencies within categorical data. Neither can be used for the given scenario. Line graphs display two numerical variables on a coordinate grid and show trends among the variables, so this is the correct choice.

67. A: First, the sample mean must be calculated.

$$\bar{x} = \frac{1}{4}(1 + 3 + 5 + 7) = 4$$

The standard deviation of the data set is:

$$\sigma = \sqrt{\frac{\Sigma(x - \bar{x})^2}{n - 1}}$$

and $n = 4$ represents the number of data points. Therefore,

$$\sigma = \sqrt{\frac{1}{3}[(1 - 4)^2 + (3 - 4)^2 + (5 - 4)^2 + (7 - 4)^2]} = \sqrt{\frac{1}{3}(9 + 1 + 1 + 9)} = 2.58$$

68. A: To calculate the range in a set of data, subtract the highest value with the lowest value. In this graph, the range of Mr. Lennon's students is 5, which can be seen physically in the graph as having the smallest difference compared with the other teachers between the highest value and the lowest value.

Language

Vocabulary Acquisition and Use

Determine or Clarify the Meaning Of Unknown and Multiple-Meaning Words and Phrases

<u>Meaning of Words and Phrases</u>
Another useful vocabulary skill is being able to understand meaning in context. A word's *context* refers to other words and information surrounding it, which can have a big impact on how readers interpret that word's meaning. Of course, many words have more than one definition. For example, consider the meaning of the word "engaged." The first definition that comes to mind might be "promised to be married," but consider the following sentences:

a. The two armies engaged in a conflict that lasted all night.

b. The three-hour lecture flew by because students were so engaged in the material.

c. The busy executive engaged a new assistant to help with his workload.

Were any of those sentences related to marriage? In fact, "engaged" has a variety of other meanings. In these sentences, respectively, it can mean: "battled," "interested or involved," and "appointed or employed." Readers may wonder how to decide which definition to apply. The appropriate meaning is prioritized based on context. For example, sentence *C* mentions "executive," "assistant," and "workload," so readers can assume that "engaged" has something to do with work—in which case, "appointed or employed" is the best definition for this context. Context clues can also be found in sentence *A*. Words like "armies" and "conflicts" show that this sentence is about a military situation, so in this context, "engaged" is closest in meaning to "battled." By using context clues—the surrounding words in the sentence—readers can easily select the most appropriate definition.

Context clues can also help readers when they don't know *any* meanings for a certain word. Test writers will deliberately ask about unfamiliar vocabulary to measure your ability to use context to make an educated guess about a word's meaning.

Which of the following is the closest in meaning to the word "loquacious" in the following sentence?

The *loquacious* professor was notorious for always taking too long to finish his lectures.
    a. knowledgeable
    b. enthusiastic
    c. approachable
    d. talkative

Even if the word "loquacious" seems completely new, it's possible to utilize context to make a good guess about the word's meaning. Grammatically, it's apparent that "loquacious" is an adjective that modifies the noun "professor"—so "loquacious" must be some kind of quality or characteristic. A clue in this sentence is "taking too long to finish his lectures." Readers should then consider qualities that might cause a professor's lectures to run long. Perhaps he's "disorganized," "slow," or "talkative"—all words that might still make sense in this sentence. Choice *D*, therefore, is a logical choice for this sentence—the professor talks too much, so his lectures run late. In fact, "loquacious" means "talkative or wordy."

One way to use context clues is to think of potential replacement words before considering the answer choices. You can also turn to the answer choices first and try to replace each of them in the sentence to see if the sentence is logical and retains the same meaning.

Another way to use context clues is to consider clues in the word itself. Most students are familiar with prefixes, suffixes, and root words—the building blocks of many English words. A little knowledge goes a long way when it comes to these components of English vocabulary, and these words can point readers in the right direction when they need help finding an appropriate definition.

Specific Word Choices on Meaning and Tone

Just as one word may have different meanings, the same meaning can be conveyed by different words or synonyms. However, there are very few synonyms that have *exactly* the same definition. Rather, there are slight nuances in usage and meaning. In this case, a writer's *diction*, or word choice, is important to the meaning meant to be conveyed.

Many words have a surface *denotation* and a deeper *connotation*. A word's *denotation* is the literal definition of a word that can be found in any dictionary (an easy way to remember this is that "denotation" and "dictionary definition" all begin with the letter "D"). For example, if someone looked up the word "snake" in the dictionary, they'd learn that a snake is a common reptile with scales, a long body, and no limbs.

A word's *connotation* refers to its emotional and cultural associations, beyond its literal definition. Some connotations are universal, some are common within a particular cultural group, and some are more personal. Let's go back to the word "snake." A reader probably already knows its denotation—a slithering animal—but readers should also take a moment to consider its possible connotations. For readers from a Judeo-Christian culture, they might associate a snake with the serpent from the Garden of Eden who tempts Adam and Eve into eating the forbidden fruit. In this case, a snake's connotations might include deceit, danger, and sneakiness.

Consider the following character description:

He slithered into the room like a snake.

Does this sound like a character who can be trusted? It's the connotation of the word "snake" that implies untrustworthiness. Connotative language, then, helps writers to communicate a deeper, more emotional meaning.

Read the following excerpt from "The Lamb," a poem by William Blake.

Little lamb, who made thee?
Dost thou know who made thee,
Gave thee life, and bid thee feed
By the stream and o'er the mead;
Gave thee clothing of delight,
Softest clothing, woolly, bright;
Gave thee such a tender voice,

230

Making all the vales rejoice?
Little lamb, who made thee?
Dost thou know who made thee?

Think about the connotations of a "lamb." Whereas a snake might make readers think of something dangerous and dishonest, a lamb tends to carry a different connotation: innocence and purity. Blake's poem contains other emotional language—"delight," "softest," "tender," "rejoice"—to support this impression.

Some words have similar denotations but very different connotations. "Weird" and "unique" can both describe something distinctive and unlike the norm. But they convey different emotions:

You have such a weird fashion sense!

You have such a unique fashion sense!

Which sentence is a compliment? Which sentence is an insult? "Weird" generally has more negative connotations, whereas "unique" is more positive. In this way, connotative language is a powerful way for writers to evoke emotion.

A writer's diction also informs their tone. *Tone* refers to the author's attitude toward their subject. A writer's tone might be critical, curious, respectful, dismissive, or any other possible attitude. The key to understanding tone is focusing not just on *what* is said, but on *how* it's said.

a. Although the latest drug trial did not produce a successful vaccine, medical researchers are one step further on the path to eradicating this deadly virus.

b. Doctors faced yet another disappointing setback in their losing battle against the killer virus; their most recent drug trial has proved as unsuccessful as the last.

Both sentences report the same information: the latest drug trial was a failure. However, each sentence presents this information in a different way, revealing the writer's tone. The first sentence has a more hopeful and confident tone, downplaying the doctors' failure ("although" it failed) and emphasizing their progress ("one step further"). The second sentence has a decidedly more pessimistic and defeatist tone, using phrases like "disappointing setback" and "losing battle." The details a writer chooses to include can also help readers to identify their attitude towards their subject matter.

Identifying emotional or connotative language can be useful in determining the tone of a text. Readers can also consider questions such as, "Who is the speaker?" or "Who is their audience?" (Remember, particularly in fiction, that the speaker or narrator may not be the same person as the author.) For example, in an article about military conflict written by a notable anti-war activist, readers might expect their tone to be critical, harsh, or cynical. If they are presented with a poem written between newlyweds, readers might expect the tone to be loving, sensitive, or infatuated. If the tone seems wildly different from what's expected, consider if the writer is using *irony*. When a writer uses irony, they say one thing but imply the opposite meaning.

Acquire and Use Accurately General Academic and Domain-Specific Words and Phrases

Vocabulary is the words a person uses on a daily basis. Having a good vocabulary is important. It's important in writing and also when you talk to people. Many of the questions on the test may have words that you don't know. Therefore, it's important to learn ways to find out a word's meaning.

It's hard to use vocabulary correctly. Imagine being thrust into a foreign country. If you didn't know right words to use to ask for the things you need, you could run into trouble! Asking for help from people who don't share the same vocabulary is hard. Language helps us understand each other. The more vocabulary words a person knows, the easier they can ask for things they need. This section of the study guide focuses on getting to know vocabulary through basic grammar.

Prefixes and Suffixes

In this section, we will look at the *meaning* of various prefixes and suffixes when added to a root word. A *prefix* is a combination of letters found at the beginning of a word. A *suffix* is a combination of letters found at the end. A *root word* is the word that comes after the prefix, before the suffix, or between them both. Sometimes a root word can stand on its own without either a prefix or a suffix.

More simply put:

Prefix + Root Word = Word

Root Word + Suffix = Word

Prefix + Root Word + Suffix = Word

Root Word = Word

Knowing the definitions of common prefixes and suffixes is helpful. It's helpful when you are trying to find out the meaning of a word you don't know. Also, knowing prefixes can help you find out the number of things, the negative of something, or the time and space of an object! Understanding suffixes can help when trying to find out the meaning of an adjective, noun, or verb.

The following charts look at some of the most common prefixes, what they mean, and how they're used to find out a word's meaning:

Number and Quantity Prefixes

Prefix	Definition	Example
bi-	two	bicycle, bilateral
mono-	one, single	monopoly, monotone
poly-	many	polygamy, polygon
semi-	half, partly	semiannual, semicircle
uni-	one	unicycle, universal

Here's an example of a number prefix:

The girl rode on a *bicycle* to school.

Look at the word *bicycle*. The root word (*cycle*)comes from the Greek and means *wheel*. The prefix *bi-* means *two*. The word *bicycle* means two wheels! When you look at any bicycles, they all have two wheels. If you had a unicycle, your bike would only have one wheel, because *uni-* means *one*.

Negative Prefixes

Prefix	Definition	Example
a-	without, lack of	amoral, atypical
in-	not, opposing	inability, inverted
non-	not	nonexistent, nonstop
un-	not, reverse	unable, unspoken

Here's an example of a negative prefix:

The girl was *insensitive* to the boy who broke his leg.

Look at the word *insensitive*. In the chart above, the prefix *in-* means *not* or *opposing*. Replace the prefix with *not*. Now place *not* in front of the word *sensitive*. Now we see that the girl was "not sensitive" to the boy who broke his leg. In simpler terms, she showed that she did not care. These are easy ways to use prefixes and suffixes in order to find out what a word means.

Time and Space Prefixes

Prefix	Definition	Example
a-	in, on, of, up, to	aloof, associate
ab-	from, away, off	abstract, absent
ad-	to, towards	adept, adjacent
ante-	before, previous	antebellum, antenna
anti-	against, opposing	anticipate, antisocial
cata-	down, away, thoroughly	catacomb, catalogue
circum-	around	circumstance, circumvent
com-	with, together, very	combine, compel
contra-	against, opposing	contraband, contrast
de-	from	decrease, descend
dia-	through, across, apart	diagram, dialect
dis-	away, off, down, not	disregard, disrespect
epi-	upon	epidemic, epiphany
ex-	out	example, exit
hypo-	under, beneath	hypoallergenic, hypothermia
inter-	among, between	intermediate, international
intra-	within	intrapersonal, intravenous
ob-	against, opposing	obtain, obscure
per-	through	permanent, persist
peri-	around	periodontal, periphery
post-	after, following	postdate, postoperative
pre-	before, previous	precede, premeditate
pro-	forward, in place of	program, propel
retro-	back, backward	retroactive, retrofit
sub-	under, beneath	submarine, substantial
super-	above, extra	superior, supersede
trans-	across, beyond, over	transform, transmit
ultra-	beyond, excessively	ultraclean, ultralight

Here's an example of a space prefix:

> The teacher's motivational speech helped *propel* her students toward greater academic achievement.

Look at the word *propel*. The prefix *pro-* means *forward*. *Forward* means something related to time and space. *Propel* means to drive or move in a forward direction. Therefore, knowing the prefix *pro-* helps interpret that the students are moving forward *toward greater academic achievement*.

Miscellaneous Prefixes

Prefix	Definition	Example
belli-	war, warlike	bellied, belligerent
bene-	well, good	benediction, beneficial
equi-	equal	equidistant, equinox
for-	away, off, from	forbidden, forsaken
fore-	previous	forecast, forebode
homo-	same, equal	homogeneous, homonym
hyper-	excessive, over	hyperextend, hyperactive
in-	in, into	insignificant, invasive
magn-	large	magnetic, magnificent
mal-	bad, poorly, not	maladapted, malnourished
mis-	bad, poorly, not	misplace, misguide
mor-	death	mortal, morgue
neo-	new	neoclassical, neonatal
omni-	all, everywhere	omnipotent, omnipresent
ortho-	right, straight	orthodontist, orthopedic
over-	above	overload, overstock,
pan-	all, entire	panacea, pander
para-	beside, beyond	paradigm, parameter
phil-	love, like	philanthropy, philosophic
prim-	first, early	primal, primer
re-	backward, again	reload, regress
sym-	with, together	symmetry, symbolize
vis-	to see	visual, visibility

Here's another prefix example:

The computer was *primitive*; it still had a floppy disk drive!

The word *primitive* has the prefix *prim-*. The prefix *prim-*indicates being *first* or *early*. *Primitive* means the early stages of evolution. It also could mean the historical development of something. Therefore, the sentence is saying that the computer is an older model, because it no longer has a floppy disk drive.

The charts that follow review some of the most common suffixes. They also include examples of how the suffixes are used to determine the meaning of a word. Remember, suffixes are added to the *end* of a root word:

Adjective Suffixes

Suffix	Definition	Example
-able (-ible)	capable of being	teachable, accessible
-esque	in the style of, like	humoresque, statuesque
-ful	filled with, marked by	helpful, deceitful
-ic	having, containing	manic, elastic
-ish	suggesting, like	malnourish, tarnish
-less	lacking, without	worthless, fearless
-ous	marked by, given to	generous, previous

Here's an example of an adjective suffix:

The live model looked so *statuesque* in the window display; she didn't even move!

Look at the word *statuesque*. The suffix *-esque* means *in the style of* or *like*. If something is *statuesque*, it's *like a statue*. In this sentence, the model looks like a statue.

Noun Suffixes

Suffix	Definition	Example
-acy	state, condition	literacy, legacy
-ance	act, condition, fact	distance, importance
-ard	one that does	leotard, billiard
-ation	action, state, result	legislation, condemnation
-dom	state, rank, condition	freedom, kingdom
-er (-or)	office, action	commuter, spectator
-ess	feminine	caress, princess
-hood	state, condition	childhood, livelihood
-ion	action, result, state	communion, position
-ism	act, manner, doctrine	capitalism, patriotism
-ist	worker, follower	stylist, activist
-ity (-ty)	state, quality, condition	community, dirty
-ment	result, action	empowerment, segment
-ness	quality, state	fitness, rudeness
-ship	position	censorship, leadership
-sion (-tion)	state, result	tension, transition
-th	act, state, quality	twentieth, wealth
-tude	quality, state, result	attitude, latitude

Look at the following example of a noun suffix:

The *spectator* cheered when his favorite soccer team scored a goal.

Look at the word *spectator*. The suffix *-or* means *action*. In this sentence, the *action* is to *spectate* (watch something). Therefore, a *spectator* is someone involved in watching something.

Verb Suffixes

Suffix	Definition	Example
-ate	having, showing	facilitate, integrate
-en	cause to be, become	frozen, written
-fy	make, cause to have	modify, rectify
-ize	cause to be, treat with	realize, sanitize

Here's an example of a verb suffix:

The preschool had to *sanitize* the toys every Tuesday and Thursday.

In the word *sanitize*, the suffix *-ize* means *cause to be* or *treat with*. By adding the suffix *-ize* to the root word *sanitary*, the meaning of the word becomes active: *cause to be sanitary*.

Text Types and Purposes

Write Arguments to Support Claims in an Analysis of Substantive Topics or Texts

Understanding the Task, Purpose, and Audience
Identifying the Task, Purpose, and Intended Audience
An author's *writing style*—the way in which words, grammar, punctuation, and sentence fluidity are used—is the most influential element in a piece of writing, and it is dependent on the purpose and the audience for whom it is intended. Together, a writing style and mode of writing form the foundation of a written work, and a good writer will choose the most effective mode and style to convey a message to readers.

Writers should first determine what they are trying to say and then choose the most effective mode of writing to communicate that message. Different writing modes and *word choices* will affect the tone of a piece—that is, its underlying attitude, emotion, or character. The argumentative mode may utilize words that are earnest, angry, passionate, or excited whereas an informative piece may have a sterile, germane, or enthusiastic tone. The tones found in narratives vary greatly, depending on the purpose of the writing. *Tone* will also be affected by the audience—teaching science to children or those who may be uninterested would be most effective with enthusiastic language and exclamation points whereas teaching science to college students may take on a more serious and professional tone, with fewer charged words and punctuation choices that are inherent to academia.

Sentence fluidity—whether sentences are long and rhythmic or short and succinct—also affects a piece of writing as it determines the way in which a piece is read. Children or audiences unfamiliar with a subject do better with short, succinct sentence structures as these break difficult concepts up into shorter points. A period, question mark, or exclamation point is literally a signal for the reader to stop and takes more time to process. Thus, longer, more complex sentences are more appropriate for adults or educated audiences as they can fit more information in between processing time.

The amount of *supporting detail* provided is also tailored to the audience. A text that introduces a new subject to its readers will focus more on broad ideas without going into greater detail whereas a text that focuses on a more specific subject is likely to provide greater detail about the ideas discussed.

Writing styles, like modes, are most effective when tailored to their audiences. Having awareness of an audience's demographic is one of the most crucial aspects of properly communicating an argument, a story, or a set of information.

Choosing the Most Appropriate Type of Writing

Before beginning any writing, it is imperative that a writer have a firm grasp on the message he or she wishes to convey and how he or she wants readers to be affected by the writing. For example, does the author want readers to be more informed about the subject? Does the writer want readers to agree with his or her opinion? Does the writer want readers to get caught up in an exciting narrative? The following steps are a guide to determining the appropriate type of writing for a task, purpose, and audience:

1. Identifying the purpose for writing the piece
2. Determining the audience
3. Adapting the writing mode, word choices, tone, and style to fit the audience and the purpose

It is important to distinguish between a work's purpose and its main idea. The essential difference between the two is that the *main idea* is what the author wants to communicate about the topic at hand whereas the *primary purpose* is why the author is writing in the first place. The primary purpose is what will determine the type of writing an author will choose to utilize, not the main idea, though the two are related. For example, if an author writes an article on the mistreatment of animals in factory farms and, at the end, suggests that people should convert to vegetarianism, the main idea is that vegetarianism would reduce the poor treatment of animals. The primary purpose is to convince the reader to stop eating animals. Since the primary purpose is to galvanize an audience into action, the author would choose the argumentative writing mode.

The next step is to consider to whom the author is appealing as this will determine the type of details to be included, the diction to be used, the tone to be employed, and the sentence structure to be used. An audience can be identified by considering the following questions:

- What is the purpose for writing the piece?
- To whom is it being written?
- What is their age range?
- Are they familiar with the material being presented, or are they just being newly introduced to it?
- Where are they from?
- Is the task at hand in a professional or casual setting?
- Is the task at hand for monetary gain?

These are just a few of the numerous considerations to keep in mind, but the main idea is to become as familiar with the audience as possible. Once the audience has been understood, the author can then adapt the writing style to align with the readers' education and interests. The audience is what determines the *rhetorical appeal* the author will use—ethos, pathos, or logos. *Ethos* is a rhetorical appeal to an audience's ethics and/or morals. Ethos is most often used in argumentative and informative writing modes. *Pathos* is an appeal to the audience's emotions and sympathies, and it is found in argumentative, descriptive, and narrative writing modes. *Logos* is an appeal to the audience's logic and reason and is used primarily in informative texts as well as in supporting details for argumentative pieces. Rhetorical appeals are discussed in depth in the informational texts and rhetoric section of the test.

If the author is trying to encourage global conversion to vegetarianism, he or she may choose to use all three rhetorical appeals to reach varying personality types. Those who are less interested in the welfare of animals but are interested in facts and science would relate more to logos. Animal lovers would relate better to an emotional appeal. In general, the most effective works utilize all three appeals.

Finally, after determining the writing mode and rhetorical appeal, the author will consider word choice, sentence structure, and tone, depending on the purpose and audience. The author may choose words that convey sadness or anger when speaking about animal welfare if writing to persuade, or he or she will stick to dispassionate and matter-of-fact tones, if informing the public on the treatment of animals in factory farms. If the author is writing to a younger or less-educated audience, he or she may choose to shorten and simplify sentence structures and word choice. If appealing to an audience with more expert knowledge on a particular subject, writers will more likely employ a style of longer sentences and more complex vocabulary.

Depending on the task, the author may choose to use a first person, second person, or third person point of view. First person and second person perspectives are inherently more casual in tone, including the author and the reader in the rhetoric, while third person perspectives are often seen in more professional settings.

Using Clear and Coherent Writing
Identifying Details that Develop a Main Idea
The main idea of a piece is the central topic or theme. To identify the main idea, a reader should consider these questions: "What's the point? What does the author want readers to take away from this text?" Everything within the selection should relate back to the main idea because it is the center of the organizational web of any written work. Particularly in articles and reports, the main idea often appears within the opening paragraphs to orient the reader to what the author wants to say about the subject. A sentence that expresses the main idea is known as a thesis sentence or *thesis statement*.

After the main idea has been introduced, *supporting details* are what develop the main idea—they make up the bulk of the work. Without supporting details, the main idea would simply be a statement, so additional details are needed to give that statement weight and validity. Supporting details can often be identified by recognizing the key words that introduce them. The following example offers several supporting details, with key words underlined:

> Man did not evolve from apes. Though we share a common ancestor, humans and apes originated through very different evolutionary paths. There are several reasons why this is true. The <u>first</u> reason is that, logically, if humans evolved from apes, modern-day apes and humans would not coexist. Evolution occurs when a genetic mutation in a species ensures the survival over the rest of the species, allowing them to pass on their genes and create a new lineage of organisms. <u>Another</u> reason is that hominid fossils only fall into one of two categories—ape-like or human-like—and there are very strong differences between the two. Australopithecines, the hominid fossils originally believed to be ancestral to humans, are ape-like, indicated by their long arms, long and curved fingers, and funnel-shaped chests. Their hand bones suggest that they "knuckle-walked" like modern day chimpanzees and gorillas, something not found in *Homo sapien* fossils. <u>Finally</u>, there is no fossilized evidence to suggest a transition between the ape-like ancestor and the *Homo sapien*, indicating that a sudden mutation may have been responsible. These and many other reasons are indicative that humans and ape-like creatures are evolutionarily different.

The underlined words—*first, another,* and *finally*—are the key words that identify the supporting details. These details can be summarized as follows:

- Humans and apes could not coexist.
- Human-like and ape-like fossils are very different.
- No transition is seen between humans and ape-like ancestors.

The supporting details all relate to the central idea that "Man did not evolve from apes," which is the first sentence of the paragraph.

Even though supporting details are more specific than the main idea, they should nevertheless all be directly related to the main idea. Without sufficient supporting details, the writer's main idea will be too weak to be effective.

Organizing a Text Clearly and Coherently

There are five basic elements inherent in effective writing, and each will be discussed throughout the various subheadings of this section.

- *Main idea*: The driving message of the writing, clearly stated or implied

- *Clear organization*: The effective and purposeful arrangement of the content to support the main idea

- *Supporting details/evidence*: Content that gives appropriate depth and weight to the main idea of the story, argument, or information

- *Diction/tone*: The type of language, vocabulary, and word choice used to express the main idea, purposefully aligned to the audience and purpose

- *Adherence to conventions of English*: Correct spelling, grammar, punctuation, and sentence structure, allowing for clear communication of ideas

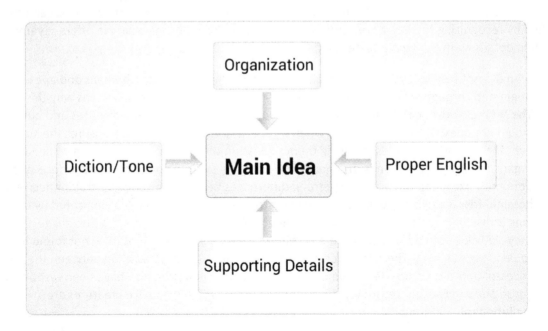

Using Varied and Effective Transitions

Transitions are the glue that holds the writing together. They function to purposefully incorporate new topics and supporting details in a smooth and coherent way. Usually, transitions are found at the beginnings of sentences, but they can also be located in the middle as a way to link clauses together. There are two types of clauses: independent and dependent as discussed in the language use and vocabulary section.

Transition words connect clauses within and between sentences for smoother writing. "I dislike apples. They taste like garbage." is choppier than "I dislike apples because they taste like garbage." Transitions demonstrate the relationship between ideas, allow for more complex sentence structures, and can alert the reader to which type of organizational format the author is using. For example, the above selection on human evolution uses the words *first, another,* and *finally* to indicate that the writer will be listing the reasons why humans and apes are evolutionarily different.

Transition words can be categorized based on the relationships they create between ideas:

- *General order*: signaling elaboration of an idea to emphasize a point—e.g., *for example, for instance, to demonstrate, including, such as, in other words, that is, in fact, also, furthermore, likewise, and, truly, so, surely, certainly, obviously, doubtless*

- *Chronological order*: referencing the time frame in which main event or idea occurs—e.g., *before, after, first, while, soon, shortly thereafter, meanwhile*

- *Numerical order/order of importance*: indicating that related ideas, supporting details, or events will be described in a sequence, possibly in order of importance—e.g., *first, second, also, finally, another, in addition, equally important, less importantly, most significantly, the main reason, last but not least*

- *Spatial order*: referring to the space and location of something or where things are located in relation to each other—e.g., *inside, outside, above, below, within, close, under, over, far, next to, adjacent to*

- *Cause and effect order*: signaling a causal relationship between events or ideas—e.g., *thus, therefore, since, resulted in, for this reason, as a result, consequently, hence, for, so*

- *Compare and contrast order*: identifying the similarities and differences between two or more objects, ideas, or lines of thought—e.g., *like, as, similarly, equally, just as, unlike, however, but, although, conversely, on the other hand, on the contrary*

- *Summary order*: indicating that a particular idea is coming to a close—e.g., *in conclusion, to sum up, in other words, ultimately, above all*

Sophisticated writing also aims to avoid overuse of transitions and ensure that those used are meaningful. Using a variety of transitions makes the writing appear more lively and informed and helps readers follow the progression of ideas.

Justifying Stylistic Choices

Stylistic choices refer to elements such as a writer's diction, sentence structure, and use of figurative language. A writer's *diction* is his or her word choice and may be elevated, academic, conversational, humorous, or any other style. The choice of diction depends on the purpose of a piece of writing. A textbook or a research paper is likely to use academic diction whereas a blog post will use more conversational expressions.

Sentence structure also affects an author's writing style. Will he or she use short, staccato sentences or longer sentences with complex structure? Effective writing tends to incorporate both styles to increase reader interest or to punctuate ideas.

Figurative language includes the use of simile, metaphor, hyperbole, or allusion, to name but a few examples. Creative or descriptive writing is likely to incorporate more non-literal expressions than academic or informative writing will. Instructors should allow students to experiment with different styles of writing and understand how style affects expression and understanding.

Introducing, Developing, and Concluding a Text Effectively

Almost all coherent written works contain three primary parts: a beginning, middle, and end. The organizational arrangements differ widely across distinct writing modes. Persuasive and expository texts utilize an introduction, body, and conclusion whereas narrative works use an orientation, series of events/conflict, and a resolution.

Every element within a written piece relates back to the main idea, and the beginning of a persuasive or expository text generally conveys the main idea or the purpose. For a narrative piece, the beginning is the section that acquaints the reader with the characters and setting, directing them to the purpose of the writing. The main idea in narrative may be implied or addressed at the end of the piece.

Depending on the primary purpose, the arrangement of the middle will adhere to one of the basic organizational structures described in the information texts and rhetoric section. They are cause and effect, problem and solution, compare and contrast, description/spatial, sequence, and order of importance.

The ending of a text is the metaphorical wrap-up of the writing. A solid ending is crucial for effective writing as it ties together loose ends, resolves the action, highlights the main points, or repeats the central idea. A conclusion ensures that readers come away from a text understanding the author's main idea. The table below highlights the important characteristics of each part of a piece of writing.

Structure	Argumentative/Informative	Narrative
Beginning	Introduction *Purpose, main idea*	Orientation *Introduces characters, setting, necessary background*
Middle	Body *Supporting details, reasons and evidence*	Events/Conflict *Story's events that revolve around a central conflict*
End	Conclusion *Highlights main points, summarizes and paraphrases ideas, reiterates the main idea*	Resolution *The solving of the central conflict*

Evaluation of Reasoning

Critical Thinking Skills

It's important to read any piece of writing critically. The goal is to discover the point and purpose of what the author is writing about through analysis. It's also crucial to establish the point or stance the author has taken on the topic of the piece. After determining the author's perspective, readers can then more effectively develop their own viewpoints on the subject of the piece.

It is important to distinguish between *fact and opinion* when reading a piece of writing. A fact is information that can be proven true. If information can be disproved, it is not a fact. For example, water freezes at or below thirty-two degrees Fahrenheit. An argument stating that water freezes at seventy degrees Fahrenheit cannot be supported by data and is therefore not a fact. Facts tend to be associated with science, mathematics, and statistics. Opinions are information open to debate. Opinions are often tied to subjective concepts like equality, morals, and rights. They can also be controversial.

Authors often use words like *think, feel, believe,* or *in my opinion* when expressing opinion, but these words won't always appear in an opinion piece, especially if it is formally written. An author's opinion may be backed up by facts, which gives it more credibility, but that opinion should not be taken as fact. A critical reader should be suspect of an author's opinion, especially if it is only supported by other opinions.

Fact	Opinion
There are 9 innings in a game of baseball.	Baseball games run too long.
James Garfield was assassinated on July 2, 1881.	James Garfield was a good president.
McDonalds has stores in 118 countries.	McDonalds has the best hamburgers.

Critical readers examine the facts used to support an author's argument. They check the facts against other sources to be sure those facts are correct. They also check the validity of the sources used to be sure those sources are credible, academic, and/or peer-reviewed. Consider that when an author uses another person's opinion to support his or her argument, even if it is an expert's opinion, it is still only an opinion and should not be taken as fact. A strong argument uses valid, measurable facts to support ideas. Even then, the reader may disagree with the argument as it may be rooted in his or her personal beliefs.

An authoritative argument may use the facts to sway the reader. In an example of global warming, many experts differ in their opinions of what alternative fuels can be used to aid in offsetting it. Because of this, a writer may choose to only use the information and expert opinion that supports his or her viewpoint.

If the argument is that wind energy is the best solution, the author will use facts that support this idea. That same author may leave out relevant facts on solar energy. The way the author uses facts can influence the reader, so it's important to consider the facts being used, how those facts are being presented, and what information might be left out.

Critical readers should also look for errors in the argument such as logical fallacies and bias. A *logical fallacy* is a flaw in the logic used to make the argument. Logical fallacies include slippery slope, straw man, and begging the question. Authors can also reflect *bias* if they ignore an opposing viewpoint or present their side in an unbalanced way. A strong argument considers the opposition and finds a way to refute it. Critical readers should look for an unfair or one-sided presentation of the argument and be skeptical, as a bias may be present. Even if this bias is unintentional, if it exists in the writing, the reader should be wary of the validity of the argument.

Readers should also look for the use of *stereotypes,* which refer to specific groups. Stereotypes are often negative connotations about a person or place and should always be avoided. When a critical reader finds stereotypes in a piece of writing, they should immediately be critical of the argument and consider the validity of anything the author presents. Stereotypes reveal a flaw in the writer's thinking and may suggest a lack of knowledge or understanding about the subject.

Identifying False Statements

A reader must also be able to identify any *logical fallacies*—logically flawed statements—that an author may make as those fallacies impact the validity and veracity of the author's claims.

Some of the more common fallacies are shown in the following chart.

Fallacy	Definition
Slippery Slope	A fallacy that is built on the idea that a particular action will lead to a series of events with negative results
Red Herring	The use of an observation or distraction to remove attention from the actual issue
Straw Man	An exaggeration or misrepresentation of an argument so that it is easier to refute
Post Hoc Ergo Propter Hoc	A fallacy that assumes an event to be the consequence of an earlier event merely because it came after it
Bandwagon	A fallacy that assumes because the majority of people feel or believe a certain way then it must be the right way
Ad Hominem	The use of a personal attack on the person or persons associated with a certain argument rather than focusing on the actual argument itself

Readers who are aware of the types of fallacious reasoning are able to weigh the credibility of the author's statements in terms of effective argument. Rhetorical text that contains a myriad of fallacious statements should be considered ineffectual and suspect.

Interpreting Textual Evidence

Literal and Figurative Meanings

It is important when evaluating texts to consider the use of both literal and figurative meanings. The words and phrases an author chooses to include in a text must be evaluated. How does the word choice affect the meaning and tone? By recognizing the use of literal and figurative language, a reader can

more readily ascertain the message or purpose of a text. Literal word choice is the easiest to analyze as it represents the usual and intended way a word or phrase is used. While figurative language is typically associated with fiction and poetry, it can be found in informational texts as well. The reader must determine not only what is meant by the figurative language in context, but also how the author intended it to shape the overall text.

Inference

Inference refers to the reader's ability to understand the unwritten text, i.e., "read between the lines" in terms of an author's intent or message. The strategy asks that a reader not take everything he or she reads at face value but instead, add his or her own interpretation of what the author seems to be trying to convey. A reader's ability to make inferences relies on his or her ability to think clearly and logically about the text. It does not ask that the reader make wild speculation or guess about the material but demands that he or she be able to come to a sound conclusion about the material.

An author's use of less literal words and phrases requires readers to make more inference when they read. Since inference involves *deduction*—deriving conclusions from ideas assumed to be true—there's more room for interpretation. Still, critical readers who employ inference, if careful in their thinking, can still arrive at the logical, sound conclusions the author intends.

Textual Evidence

Once a reader has determined an author's thesis or main idea, he or she will need to understand how textual evidence supports interpretation of that thesis or main idea. Test takers will be asked direct questions regarding an author's main idea and may be asked to identify evidence that would support those ideas. This will require test takers to comprehend literal and figurative meanings within the text passage, be able to draw inferences from provided information, and be able to separate important evidence from minor supporting detail. It's often helpful to skim test questions and answer options prior to critically reading; however, test takers should avoid the temptation to solely look for the correct answers. Just trying to find the "right answer" may cause test takers to miss important supporting textual evidence. Making mental note of test questions is only helpful as a guide when reading.

After identifying an author's thesis or main idea, a test taker should look at the supporting details that the author provides to back up his or her assertions, identifying those additional pieces of information that help expand the thesis. From there, test takers should examine the additional information and related details for credibility, the author's use of outside sources, and be able to point to direct evidence that supports the author's claims. It's also imperative that test takers be able to identify what is strong support and what is merely additional information that is nice to know but not necessary. Being able to make this differentiation will help test takers effectively answer questions regarding an author's use of supporting evidence.

Write Informative/Explanatory Texts to Examine and Convey Ideas, Concepts, and Information Clearly and Accurately

Informational texts are a category of texts within the genre of nonfiction. Their intent is to inform, and while they do convey a point of view and may include literary devices, they do not utilize other literary elements, such as characters or plot. An informational text also reflects a *thesis*—an implicit or explicit statement of the text's intent and/or a *main idea*—the overarching focus and/or purpose of the text, generally implied. Some examples of informational texts are informative articles, instructional/how-to texts, factual reports, reference texts, and self-help texts.

Understanding Organizational Patterns and Structures

Organizational Structure within Informational Text

Informational text is specifically designed to relate factual information, and although it is open to a reader's interpretation and application of the facts, the structure of the presentation is carefully designed to lead the reader to a particular conclusion or central idea. When reading informational text, it is important that readers are able to understand its organizational structure as the structure often directly relates to an author's intent to inform and/or persuade the reader.

The first step in identifying the text's structure is to determine the thesis or main idea. The thesis statement and organization of a work are closely intertwined. *A thesis statement* indicates the writer's purpose and may include the scope and direction of the text. It may be presented at the beginning of a text or at the end, and it may be explicit or implicit.

Once a reader has a grasp of the thesis or main idea of the text, he or she can better determine its organizational structure. Test takers are advised to read informational text passages more than once in order to comprehend the material fully. It is also helpful to examine any text features present in the text including the table of contents, index, glossary, headings, footnotes, and visuals. The analysis of these features and the information presented within them, can offer additional clues about the central idea and structure of a text. The following questions should be asked when considering structure:

- How does the author assemble the parts to make an effective whole argument?
- Is the passage linear in nature and if so, what is the timeline or thread of logic?
- What is the presented order of events, facts, or arguments? Are these effective in contributing to the author's thesis?
- How can the passage be divided into sections? How are they related to each other and to the main idea or thesis?
- What key terms are used to indicate the organization?

Next, test takers should skim the passage, noting the first line or two of each body paragraph—the *topic sentences*—and the conclusion. Key *transitional terms*, such as *on the other hand*, *also*, *because*, *however*, *therefore*, *most importantly*, and *first*, within the text can also signal organizational structure. Based on these clues, readers should then be able to identify what type of organizational structure is being used. The following organizational structures are most common:

- *Problem/solution*—organized by an analysis/overview of a problem, followed by potential solution(s)

- *Cause/effect*—organized by the effects resulting from a cause or the cause(s) of a particular effect

- *Spatial order*—organized by points that suggest location or direction—e.g., top to bottom, right to left, outside to inside

- *Chronological/sequence order*—organized by points presented to indicate a passage of time or through purposeful steps/stages

- *Comparison/Contrast*—organized by points that indicate similarities and/or differences between two things or concepts

- *Order of importance*—organized by priority of points, often most significant to least significant or vice versa

Identifying Rhetorical Strategies

Rhetoric refers to an author's use of particular strategies, appeals, and devices to persuade an intended audience. The more effective the use of rhetoric, the more likely the audience will be persuaded.

Determining an Author's Point of View

A *rhetorical strategy*—also referred to as a *rhetorical mode*—is the structural way an author chooses to present his/her argument. Though the terms noted below are similar to the organizational structures noted earlier, these strategies do not imply that the entire text follows the approach. For example, a cause and effect organizational structure is solely that, nothing more. A persuasive text may use cause and effect as a strategy to convey a singular point. Thus, an argument may include several of the strategies as the author strives to convince his or her audience to take action or accept a different point of view. It's important that readers are able to identify an author's thesis and position on the topic in order to be able to identify the careful construction through which the author speaks to the reader. The following are some of the more common rhetorical strategies:

- *Cause and effect*—establishing a logical correlation or causation between two ideas
- *Classification/division*—the grouping of similar items together or division of something into parts
- *Comparison/contrast*—the distinguishing of similarities/differences to expand on an idea
- *Definition*—used to clarify abstract ideas, unfamiliar concepts, or to distinguish one idea from another
- *Description*—use of vivid imagery, active verbs, and clear adjectives to explain ideas
- *Exemplification*—the use of examples to explain an idea
- *Narration*—anecdotes or personal experience to present or expand on a concept
- *Problem/Solution*—presentation of a problem or problems, followed by proposed solution(s)

Rhetorical Strategies and Devices

A *rhetorical device* is the phrasing and presentation of an idea that reinforces and emphasizes a point in an argument. A rhetorical device is often quite memorable. One of the more famous uses of a rhetorical device is in John F. Kennedy's 1961 inaugural address: "Ask not what your country can do for you, ask what you can do for your country." The contrast of ideas presented in the phrasing is an example of the rhetorical device of antimetabole. Some other common examples are listed on the following page, but test takers should be aware that this is not a complete list.

Device	Definition	Example
Allusion	A reference to a famous person, event, or significant literary text as a form of significant comparison	"We are apt to shut our eyes against a painful truth, and listen to the song of that siren till she transforms us into beasts." Patrick Henry
Anaphora	The repetition of the same words at the beginning of successive words, phrases, or clauses, designed to emphasize an idea	"We shall not flag or fail. We shall go on to the end. We shall fight in France, we shall fight on the seas and oceans, we shall fight with growing confidence ... we shall fight in the fields and in the streets, we shall fight in the hills. We shall never surrender." Winston Churchill
Understatement	A statement meant to portray a situation as less important than it actually is to create an ironic effect	"The war in the Pacific has not necessarily developed in Japan's favor." Emperor Hirohito, surrendering Japan in World War II
Parallelism	A syntactical similarity in a structure or series of structures used for impact of an idea, making it memorable	"A penny saved is a penny earned." Ben Franklin
Rhetorical question	A question posed that is not answered by the writer though there is a desired response, most often designed to emphasize a point	"Can anyone look at our reduced standing in the world today and say, 'Let's have four more years of this?'" Ronald Reagan

Understanding Methods Used to Appeal to a Specific Audience

Rhetorical Appeals

In an argument or persuasive text, an author will strive to sway readers to an opinion or conclusion. To be effective, an author must consider his or her intended audience. Although an author may write text for a general audience, he or she will use methods of appeal or persuasion to convince that audience. Aristotle asserted that there were three methods or modes by which a person could be persuaded. These are referred to as *rhetorical appeals*.

The three main types of rhetorical appeals are shown in the following graphic.

Ethos, also referred to as an *ethical appeal*, is an appeal to the audience's perception of the writer as credible (or not), based on their examination of their ethics and who the writer is, his/her experience or incorporation of relevant information, or his/her argument. For example, authors may present testimonials to bolster their arguments. The reader who critically examines the veracity of the testimonials and the credibility of those giving the testimony will be able to determine if the author's use of testimony is valid to his or her argument. In turn, this will help the reader determine if the author's thesis is valid. An author's careful and appropriate use of technical language can create an overall knowledgeable effect and, in turn, act as a convincing vehicle when it comes to credibility. Overuse of technical language, however, may create confusion in readers and obscure an author's overall intent.

Pathos, also referred to as a *pathetic* or *emotional appeal*, is an appeal to the audience's sense of identity, self-interest, or emotions. A critical reader will notice when the author is appealing to pathos through anecdotes and descriptions that elicit an emotion such as anger or pity. Readers should also beware of factual information that uses generalization to appeal to the emotions. While it's tempting to believe an author is the source of truth in his or her text, an author who presents factual information as universally true, consistent throughout time, and common to all groups is using *generalization*. Authors who exclusively use generalizations without specific facts and credible sourcing are attempting to sway readers solely through emotion.

Logos, also referred to as a *logical appeal*, is an appeal to the audience's ability to see and understand the logic in a claim offered by the writer. A critical reader has to be able to evaluate an author's arguments for validity of reasoning and for sufficiency when it comes to argument.

Understanding Development of a Written Argument

Evaluating an Author's Purpose

A reader must be able to evaluate the argument or point the author is trying to make and determine if it is adequately supported. The first step is to determine the main idea. The main idea is what the author wants to say about a specific topic. The next step is to locate the supporting details. An author uses supporting details to illustrate the main idea. These are the details that provide evidence or examples to help make a point. Supporting details often appear in the form of quotations, paraphrasing, or analysis. Test takers should then examine the text to make sure the author connects details and analysis to the main point. These steps are crucial to understanding the text and evaluating how well the author presents his or her argument and evidence. The following graphic demonstrates the connection between the main idea and the supporting details.

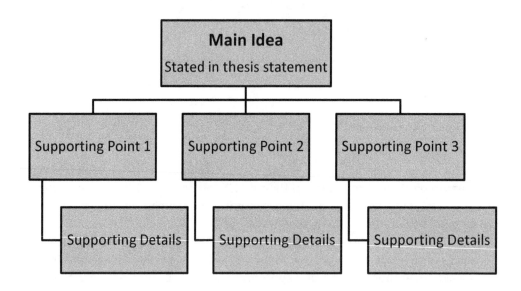

Evaluating Evidence

It is important to evaluate the author's supporting details to be sure that they are credible, provide evidence of the author's point, and directly support the main idea. Critical readers examine the facts used to support an author's argument and check those facts against other sources to be sure the facts are correct. They also check the validity of the sources used to be sure those sources are credible, academic, and/or peer- reviewed. A strong argument uses valid, measurable facts to support ideas.

Electronic Resources

With print texts, it is easy to identify the authors and their credentials, as well as the publisher and their reputation. With electronic resources like websites, though, it can be trickier to assess the reliability of information. Students should be alert when gathering information from the Internet. Understanding the significance of website *domains*—which include identification strings of a site—can help. Website domains ending in *.edu* are educational sites and tend to offer more reliable research in their field. A *.org* ending tends to be used by nonprofit organizations and other community groups, *.com* indicates a privately-owned website, and a *.gov* site is run by the government. Websites affiliated with official organizations, research groups, or institutes of learning are more likely to offer relevant, fact-checked, and reliable information.

Write Arguments Focused On Discipline-Specific Content

When it comes to writing discipline-specific content, the key is in understanding the characteristics of the specific discipline you are planning on writing about. Whether you are writing essays in high school, taking a broad range of college courses, or writing to communicate in a new career, there are many different disciplines that you will have to deal with. These disciplines are most likely one of the following:

- Business and Communication
- Humanities
- Social Sciences
- Sciences

It's important to remember that your basic five-paragraph essay that you write in high school won't cut it if you're expected to write a twenty-five-page proposal in your business class in college. Learning the expectations of the discipline you are writing for is key if in order to be successful in your writing. Here are some basic characteristics of each of the above disciplines:

Business and Communication

Business and communication writing, focuses on persuasive writing as well as informational documents. Although it's important to identify your audience in all writing situations, it is especially important in this discipline to create a list before you start writing of who will receive the document, what their status of knowledge or authority is on the subject, and how they are expected to receive the document you plan on writing. Will they be hostile to the information? If so, it's important to begin with a neutral stance. Are they on board with a business proposal you are writing? If so, focus on how to most efficiently get to the facts of what it will take to make the proposal work. Knowing your audience is key in this kind of writing. Also, it's important to pay attention to the length and organization of the document. Are they a busy executive? If so, writing an abstract to sum up the entirety of the proposal is a great idea. For organization, if the document is some kind of proposal, the most common structure to write in would be the problem-solution structure.

Humanities

Humanities is a broad field, containing various formats and styles you will have to choose from. If you are writing an essay for an English class, most likely this will be structured as an essay with a clear introduction containing a thesis statement, body paragraphs, and a conclusion regardless of whether it's a cause and effect essay or a comparison/contrast essay. Writing arguments are important in this genre, as rhetoric focuses on persuasive techniques such as audience persuasion using ethos, pathos, and logos. Sometimes the humanities involve creative works, in which organization and formatting will be up to you to decide. Whatever you are being asked to write, it's important to either research the topic and find credible sources, or ask your professor or mentor for examples of this type of writing.

Social Sciences

The social sciences field contains disciplines such as anthropology, sociology, history, and political science, among others. These disciplines are similar to the humanities in their organization; they will want straightforward writing containing an introduction, body paragraphs, and a conclusion. The social sciences are also dedicated to clear, informative writing. It's important to keep in mind to approach social science topics from a non-biased, objective tone, rather than a persuasive, personal tone that you

Test Prep Books

might take up in a humanities paper. Giving the most up-to-date information possible based on your content-base will establish you as a credible author and allow your readers to trust you.

Sciences
Writing in the sciences is much like other writing; it's important to identify your audience, present a clear, concise understanding of your topic, and to also remain objective in your findings. Science writing includes lab reports, peer-reviewed journal articles, grant proposals, and literature review articles. Credibility is extremely important in scientific writing, so it's imperative to use the correct styling and format when dealing with science writing. Most of the time, science writing uses APA style for publications, so it might be helpful to do research to find out exactly what APA-style writing entails.

Style Manual
A *style manual* is a comprehensive collection of guidelines for language use and document formatting. Some fields refer to a common style guide—e.g., the Associated Press or *AP Stylebook*, a standard in American journalism. Individual organizations may rely on their own house style. Regardless, the purpose of a style manual is to ensure uniformity across all documents. Style manuals explain things such as how to format titles, when to write out numbers or use numerals, and how to cite sources. Because there are many different style guides, students should know how and when to consult an appropriate guide. The Chicago Manual of Style is common in the publication of books and academic journals. The Modern Language Association style (MLA) is another commonly used academic style format, while the American Psychological Association style (APA) may be used for scientific publications. Familiarity with using a style guide is particularly important for students who are college bound or pursuing careers in academic or professional writing.

Write Informative/Explanatory Texts, Including the Narration of Historical Events, Scientific Procedures/Experiments, or Technical Processes

Writing informative or explanatory texts is helpful for anyone going into the social sciences, science, or technical field. As stated above, precision and conciseness are very important in this type of writing. Identifying the audience before the writing begins is also a helpful tip to guiding your writing toward the appropriate tone. Some examples of types of writing we will look at is narration of historical events, scientific procedures, and technical processes.

Narration of Historical Events
A trend in historical writing is writing in a narrative fashion about an event that happened in the past rather than an analytical document about a historical event. Historical narratives are either written in a traditional or modern narrative. Traditional narratives tend to focus more on the chronology of the event itself and does not pull in other contexts associated with that event. For example, a historical narrative written in the traditional format of the American Revolution would focus solely on events that happened within that world. Modern narratives focus more on causes of the event and are not afraid to bring in outside contexts to make sense of the event that is being narrated. For example, a modern form of the historical narrative dealing with the American Revolution may look at other revolutions around the same time as America in order to make sense of the collective social consciousness during that era.

Scientific Procedures/Experiments
Science procedures have specific directions regarding style and format. Most science procedures are written in APA-style format. This gives specific directions on how to write in-text citations, as well as

252

what references will look like when the report is complete. Scientific procedures are also written in third person past tense. Third person means an objective tone, like the following:

An experiment was conducted to see if helium balloons would rise in a zero-gravity chamber.

Rather than

We did an experiment to see if helium balloons would rise in a zero-gravity chamber.

The first example is written in third person, while the second example is written in first person, because it uses "we."

Scientific procedures are also specific in that they usually include a set of steps you must adhere to. These include determining a problem, creating a hypothesis, designating what materials to use, explaining the procedure, recording the results, returning to the hypothesis, including any errors, and concluding with the analysis of the findings.

Technical Processes

Instructional and process reports explain how a thing is done. These documents are not limited only to technical fields. Your history professor can ask for a process report regarding the chronology of wars, or your writing professor can ask for a process report over how theory evolved over a period of time. Process reports may contain descriptive or prescriptive processes. Descriptive processes describe how something is done so that the audience can better grasp the content. Prescriptive processes are instructional, and they explain how something *should* be done so that someone else can read it and do it themselves. As in any other writing, it's important for process writing to be clear. Process writing is usually written in chronological order and has images or videos to accompany the material so that visual learners can understand the process as well.

Conventions of Standard English

Demonstrate Command of the Conventions of Standard English Grammar and Usage When Writing or Speaking

Sentence Structure
Sentence Types
There are four ways in which we can structure sentences: simple, compound, complex, and compound-complex. Sentences can be composed of just one clause or many clauses joined together.

When a sentence is composed of just one clause (an independent clause), we call it a simple sentence. Simple sentences do not necessarily have to be short sentences. They just require one independent clause with a subject and a predicate. For example:

Thomas marched over to Andrew's house.

Jonah and Mary constructed a simplified version of the Eiffel Tower with Legos.

When a sentence has two or more independent clauses we call it a compound sentence. The clauses are connected by a comma and a coordinating conjunction—*and, but, or, nor, for*—or by a semicolon. Compound sentences do not have dependent clauses. For example:

> We went to the fireworks stand, and we bought enough fireworks to last all night.

> The children sat on the grass, and then we lit the fireworks one at a time.

When a sentence has just one independent clause and includes one or more dependent clauses, we call it a complex sentence:

> Because she slept well and drank coffee, Sarah was quite productive at work.

> Although Will had coffee, he made mistakes while using the photocopier.

When a sentence has two or more independent clauses and at least one dependent clause, we call it a compound-complex sentence:

> It may come as a surprise, but I found the tickets, and you can go to the show.

> Jade is the girl who dove from the high-dive, and she stunned the audience silent.

Sentence Fragments

Remember that a complete sentence must have both a subject and a verb. Complete sentences consist of at least one independent clause. Incomplete sentences are called sentence fragments. A sentence fragment is a common error in writing. Sentence fragments can be independent clauses that start with subordinating words, such as *but, as, so that,* or *because,* or they could simply be missing a subject or verb.

You can correct a fragment error by adding the fragment to a nearby sentence or by adding or removing words to make it an independent clause. For example:

> Dogs are my favorite animals. Because cats are too independent. (Incorrect; the word *because* creates a sentence fragment)

> Dogs are my favorite animals because cats are too independent. (Correct; the fragment becomes a dependent clause.)

> Dogs are my favorite animals. Cats are too independent. (Correct; the fragment becomes a simple sentence.)

Run-on Sentences

Another common mistake in writing is the run-on sentence. A run-on is created when two or more independent clauses are joined without the use of a conjunction, a semicolon, a colon, or a dash. We don't want to use commas where periods belong. Here is an example of a run-on sentence:

> Making wedding cakes can take many hours I am very impatient, I want to see them completed right away.

There are a variety of ways to correct a run-on sentence. The method you choose will depend on the context of the sentence and how it fits with neighboring sentences:

Making wedding cakes can take many hours. I am very impatient. I want to see them completed right away. (Use periods to create more than one sentence.)

Making wedding cakes can take many hours; I am very impatient—I want to see them completed right away. (Correct the sentence using a semicolon, colon, or dash.)

Making wedding cakes can take many hours and I am very impatient, so I want to see them completed right away. (Correct the sentence using coordinating conjunctions.)

I am very impatient because I would rather see completed wedding cakes right away than wait for it to take many hours. (Correct the sentence by revising.)

Dangling and Misplaced Modifiers

A modifier is a word or phrase meant to describe or clarify another word in the sentence. When a sentence has a modifier but is missing the word it describes or clarifies, it's an error called a dangling modifier. We can fix the sentence by revising to include the word that is being modified. Consider the following examples with the modifier italicized:

Having walked five miles, this bench will be the place to rest. (Incorrect; this version of the sentence implies that the bench walked the miles, not the person.)

Having walked five miles, Matt will rest on this bench. (Correct; in this version, *having walked five miles* correctly modifies *Matt*, who did the walking.)

Since midnight, my dreams have been pleasant and comforting. (Incorrect; in this version, the adverb clause *since midnight* cannot modify the noun *dreams*.)

Since midnight, I have had pleasant and comforting dreams. (Correct; in this version, *since midnight* modifies the verb *have had*, telling us when the dreams occurred.)

Sometimes the modifier is not located close enough to the word it modifies for the sentence to be clearly understood. In this case, we call the error a misplaced modifier. Here is an example with the modifier italicized and the modified word in underlined.

We gave the hot <u>cocoa</u> to the children *that was filled with marshmallows.* (Incorrect; this sentence implies that the children are what are filled with marshmallows.)

We gave the hot <u>*cocoa that was filled with marshmallows*</u> to the children. (Correct; here, the cocoa is filled with marshmallows. The modifier is near the word it modifies.)

Parallelism and Subordination

Parallelism

To be grammatically correct we must use articles, prepositions, infinitives, and introductory words for dependent clauses consistently throughout a sentence. This is called parallelism. We use parallelism when we are matching parts of speech, phrases, or clauses with another part of the sentence. Being inconsistent creates confusion. Consider the following example.

Incorrect: Be ready for running and to ride a bike during the triathlon.

Correct: Be ready to run and to ride a bike during the triathlon.

Correct: Be ready for running and for riding a bike during the triathlon.

In the incorrect example, the gerund *running* does not match with the infinitive *to ride*. Either both should be infinitives or both should be gerunds.

Subordination

Sometimes we have unequal pieces of information in a sentence where one piece is more important than the other. We need to show that one piece of information is subordinate to the other. We can make the more important piece an independent clause and connect the other piece by making it a dependent clause. Consider this example:

Central thought: Kittens can always find their mother.

Subordinate: Kittens are blind at birth.

Complex Sentence: Despite being blind at birth, kittens can always find their mother.

The sentence "Kittens are blind at birth" is made subordinate to the sentence "Kittens can always find their mother" by placing the word "Despite" at the beginning and removing the subject, thus turning an independent clause ("kittens are blind at birth") into a subordinate phrase ("Despite being blind at birth").

Usage
Nouns
A noun is a person, place, thing ,or idea. All nouns fit into one of two types, common or proper.

A *common noun* is a word that identifies any of a class of people, places, or things. Examples include numbers, objects, animals, feelings, concepts, qualities, and actions. *A, an,* or *the* usually precedes the common noun. These parts of speech are called *articles*. Here are some examples of sentences using nouns preceded by articles.

A building is under construction.
The girl would like to move to *the* city.

A *proper noun* (also called a *proper name*) is used for the specific name of an individual person, place, or organization. The first letter in a proper noun is capitalized. "My name is *Mary*." "I work for *Walmart*."

Nouns sometimes serve as adjectives (which themselves describe nouns), such as "hockey player" and "state government."

An abstract noun is an idea, state, or quality. It is something that can't be touched, such as happiness, courage, evil, or humor.

A concrete noun is something that can be experienced through the senses (touch, taste, hear, smell, see). Examples of concrete nouns are birds, skateboard, pie, and car.

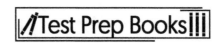

A collective noun refers to a collection of people, places, or things that act as one. Examples of collective nouns are as follows: team, class, jury, family, audience, and flock.

Pronouns

A word used in place of a noun is known as a *pronoun*. Pronouns are words like *I, mine, hers,* and *us.*

Pronouns can be split into different classifications (seen below) which make them easier to learn; however, it's not important to memorize the classifications.

- Personal pronouns: refer to people

- First person: we, I, our, mine

- Second person: you, yours

- Third person: he, them

- Possessive pronouns: demonstrate ownership (mine, my, his, yours)

- Interrogative pronouns: ask questions (what, which, who, whom, whose)

- Relative pronouns: include the five interrogative pronouns and others that are relative (whoever, whomever, that, when, where)

- Demonstrative pronouns: replace something specific (this, that, those, these)

- Reciprocal pronouns: indicate something was done or given in return (each other, one another)

- Indefinite pronouns: have a nonspecific status (anybody, whoever, someone, everybody, somebody)

Indefinite pronouns such as *anybody, whoever, someone, everybody,* and *somebody* command a singular verb form, but others such as *all, none,* and *some* could require a singular or plural verb form.

Antecedents

An *antecedent* is the noun to which a pronoun refers; it needs to be written or spoken before the pronoun is used. For many pronouns, antecedents are imperative for clarity. In particular, many of the personal, possessive, and demonstrative pronouns need antecedents. Otherwise, it would be unclear who or what someone is referring to when they use a pronoun like *he* or *this.*

Pronoun reference means that the pronoun should refer clearly to one, clear, unmistakable noun (the antecedent).

Pronoun-antecedent agreement refers to the need for the antecedent and the corresponding pronoun to agree in gender, person, and number. Here are some examples:

> The *kidneys* (plural antecedent) are part of the urinary system. *They* (plural pronoun) serve several roles."

> The kidneys are part of the *urinary system* (singular antecedent). *It* (singular pronoun) is also known as the renal system.

Pronoun Cases

The subjective pronouns —*I, you, he/she/it, we, they,* and *who*—are the subjects of the sentence.

Example: *They* have a new house.

The objective pronouns—*me, you* (*singular*), *him/her, us, them,* and *whom*—are used when something is being done for or given to someone; they are objects of the action.

Example: The teacher has an apple for *us.*

The possessive pronouns—*mine, my, your, yours, his, hers, its, their, theirs, our,* and *ours*—are used to denote that something (or someone) belongs to someone (or something).

Example: It's *their* chocolate cake.
Even Better Example: It's *my* chocolate cake!

One of the greatest challenges and worst abuses of pronouns concerns *who* and *whom.* Just knowing the following rule can eliminate confusion. *Who* is a subjective-case pronoun used only as a subject or subject complement. *Whom* is only objective-case and, therefore, the object of the verb or preposition.

Who is going to the concert?

You are going to the concert with *whom*?

Hint: When using *who* or *whom,* think of whether someone would say *he* or *him.* If the answer is *he,* use *who.* If the answer is *him,* use *whom.* This trick is easy to remember because *he* and *who* both end in vowels, and *him* and *whom* both end in the letter *M.*

Verbs

The *verb* is the part of speech that describes an action, state of being, or occurrence.

A *verb* forms the main part of a predicate of a sentence. This means that the verb explains what the noun (which will be discussed shortly) is doing. A simple example is *time flies.* The verb *flies* explains what the action of the noun, *time,* is doing. This example is a *main* verb.

Helping (*auxiliary*) verbs are words like *have, do, be, can, may, should, must,* and *will.* "I *should* go to the store." Helping verbs assist main verbs in expressing tense, ability, possibility, permission, or obligation.

Particles are minor function words like *not, in, out, up,* or *down* that become part of the verb itself. "I might *not.*"

Participles are words formed from verbs that are often used to modify a noun, noun phrase, verb, or verb phrase.

The *running* teenager collided with the cyclist.

Participles can also create compound verb forms.

He is *speaking.*

Verbs have five basic forms: the *base* form, the *-s* form, the *-ing* form, the *past* form, and the *past participle* form.

The *past* forms are either *regular* (*love/loved; hate/hated*) or *irregular* because they don't end by adding the common past tense suffix "-ed" (*go/went; fall/fell; set/set*).

Verb Forms

Shifting verb forms entails *conjugation*, which is used to indicate *tense, voice,* or *mood*.

Verb tense is used to show when the action in the sentence took place. There are several different verb tenses, and it is important to know how and when to use them. Some verb tenses can be achieved by changing the form of the verb, while others require the use of helping verbs (e.g., *is, was,* or *has*).

- Present tense shows the action is happening currently or is ongoing:

    I walk to work every morning.

    She is stressed about the deadline.

- Past tense shows that the action happened in the past or that the state of being is in the past:

    I walked to work yesterday morning.

    She was stressed about the deadline.

- Future tense shows that the action will happen in the future or is a future state of being:

    I will walk to work tomorrow morning.

    She will be stressed about the deadline.

- Present perfect tense shows action that began in the past, but continues into the present:

    I have walked to work all week.

    She has been stressed about the deadline.

- Past perfect tense shows an action was finished before another took place:

    I had walked all week until I sprained my ankle.

    She had been stressed about the deadline until we talked about it.

- Future perfect tense shows an action that will be completed at some point in the future:

    By the time the bus arrives, I will have walked to work already.

Voice

Verbs can be in the active or passive voice. When the subject completes the action, the verb is in *active voice*. When the subject receives the action of the sentence, the verb is in *passive voice*.

Active: Jamie ate the ice cream.

Passive: The ice cream was eaten by Jamie.

In active voice, the subject (*Jamie*) is the "do-er" of the action (*ate*). In passive voice, the subject *ice cream* receives the action of being eaten.

While passive voice can add variety to writing, active voice is the generally preferred sentence structure.

Mood

Mood is used to show the speaker's feelings about the subject matter. In English, there is *indicative mood, imperative mood,* and *subjective mood.*

Indicative mood is used to state facts, ask questions, or state opinions:

> Bob will make the trip next week.

> When can Bob make the trip?

Imperative mood is used to state a command or make a request:

> Wait in the lobby.

> Please call me next week.

Subjunctive mood is used to express a wish, an opinion, or a hope that is contrary to fact:

> If I were in charge, none of this would have happened.

> Allison wished she could take the exam over again when she saw her score.

Adjectives

Adjectives are words used to modify nouns and pronouns. They can be used alone or in a series and are used to further define or describe the nouns they modify.

> Mark made us a delicious, four-course meal.

The words *delicious* and *four-course* are adjectives that describe the kind of meal Mark made.

Articles are also considered adjectives because they help to describe nouns. Articles can be general or specific. The three articles in English are: a, an, and the.

Indefinite articles (a, an) are used to refer to nonspecific nouns. The article *a* proceeds words beginning with consonant sounds, and the article *an* proceeds words beginning with vowel sounds.

> A car drove by our house.

> An alligator was loose at the zoo.

> He has always wanted a ukulele. (The first *u* makes a *y* sound.)

Note that *a* and *an* should only proceed nonspecific nouns that are also singular. If a nonspecific noun is plural, it does not need a preceding article.

> Alligators were loose at the zoo.

The *definite article (the)* is used to refer to specific nouns:

> The car pulled into our driveway.

Note that *the* should proceed all specific nouns regardless of whether they are singular or plural.

> The cars pulled into our driveway.

Comparative adjectives are used to compare nouns. When they are used in this way, they take on positive, comparative, or superlative form.

> The positive form is the normal form of the adjective:

>> Alicia is tall.

> The *comparative* form shows a comparison between two things:

>> Alicia is taller than Maria.

> *Superlative* form shows comparison between more than two things:

>> Alicia is the tallest girl in her class.

Usually, the comparative and superlative can be made by adding *–er* and *–est* to the positive form, but some verbs call for the helping verbs *more* or *most*. Other exceptions to the rule include adjectives like *bad*, which uses the comparative *worse* and the superlative *worst*.

An adjective phrase is not a bunch of adjectives strung together, but a group of words that describes a noun or pronoun and, thus, functions as an adjective. *Very ugly* is an adjective phrase; so are *way too fat* and *faster than a speeding bullet.*

Adverbs

Adverbs have more functions than adjectives because they modify or qualify verbs, adjectives, or other adverbs as well as word groups that express a relation of place, time, circumstance, or cause. Therefore, adverbs answer any of the following questions: *How, when, where, why, in what way, how often, how much, in what condition,* and/or *to what degree. How good looking is he? He is <u>very</u> handsome.*

Here are some examples of adverbs for different situations:

- how: quickly
- when: daily
- where: there
- in what way: easily
- how often: often
- how much: much
- in what condition: badly
- what degree: hardly

As one can see, for some reason, many adverbs end in *-ly.*

Adverbs do things like emphasize (*really, simply,* and *so*), amplify (*heartily, completely,* and *positively*), and tone down (*almost, somewhat,* and *mildly*).

Adverbs also come in phrases.

The dog ran as <u>though his life depended on it.</u>

Prepositions

Prepositions are connecting words and, while there are only about 150 of them, they are used more often than any other individual groups of words. They describe relationships between other words. They are placed before a noun or pronoun, forming a phrase that modifies another word in the sentence. *Prepositional phrases* begin with a preposition and end with a noun or pronoun, the *object of the preposition*. *A pristine lake is <u>near the store</u> and <u>behind the bank</u>.*

Some commonly used prepositions are *about, after, anti, around, as, at, behind, beside, by, for, from, in, into, of, off, on, to,* and *with.*

Complex prepositions, which also come before a noun or pronoun, consist of two or three words such as *according to, in regards to,* and *because of.*

Conjunctions

Conjunctions are vital words that connect words, phrases, thoughts, and ideas. Conjunctions show relationships between components. There are two types:

Coordinating conjunctions are the primary class of conjunctions placed between words, phrases, clauses, and sentences that are of equal grammatical rank; the coordinating conjunctions are for, and, nor, but, or, yes, and so. A useful memorization trick is to remember that the first letter of these conjunctions collectively spell the word *fanboys*.

> I need to go shopping, *but* I must be careful to leave enough money in the bank.
> She wore a black, red, *and* white shirt.

Subordinating conjunctions are the secondary class of conjunctions. They connect two unequal parts, one *main* (or *independent*) and the other *subordinate* (or *dependent*). I must go to the store *even though* I do not have enough money in the bank.

> *Because* I read the review, I do not want to go to the movie.

Notice that the presence of subordinating conjunctions makes clauses dependent. *I read the review* is an independent clause, but *because* makes the clause dependent. Thus, it needs an independent clause to complete the sentence.

Interjections

Interjections are words used to express emotion. Examples include *wow, ouch,* and *hooray.* Interjections are often separate from sentences; in those cases, the interjection is directly followed by an exclamation point. In other cases, the interjection is included in a sentence and followed by a comma. The punctuation plays a big role in the intensity of the emotion that the interjection is expressing. Using a comma or semicolon indicates less excitement than using an exclamation mark.

Subjects

Every sentence must include a subject and a verb. The *subject* of a sentence is who or what the sentence is about. It's often directly stated and can be determined by asking "Who?" or "What?" did the action:

Most sentences contain a direct subject, in which the subject is mentioned in the sentence.

> *Kelly mowed the lawn.*

> Who mowed the lawn? *Kelly*

> *The air-conditioner ran all night*

> What ran all night? *the air-conditioner*

The subject of imperative sentences is *you*, because imperative subjects are commands. the subject is implied because it is a command:

> *Go home after the meeting*.

> Who should go home after the meeting? *you* (implied)

In *expletive sentences* that start with "there are" or "there is," the subject is found after the predicate. The subject cannot be "there," so it must be another word in the sentence:

> *There is a cup sitting on the coffee table.*

> What is sitting on the coffee table? *a cup*

Simple and Complete Subjects

A *complete subject* includes the simple subject and all the words modifying it, including articles and adjectives. A *simple subject* is the single noun without its modifiers.

> A warm, chocolate-chip cookie sat on the kitchen table.

> Complete subject: *a warm, chocolate-chip cookie*

> Simple subject: *cookie*

The words *a, warm, chocolate,* and *chip* all modify the simple subject *cookie*.

There might also be a *compound subject*, which would be two or more nouns without the modifiers.

> A little girl and her mother walked into the shop.

> Complete subject: *A little girl and her mother*

> Compound subject: *girl, mother*

In this case, *the girl and her mother* are both completing the action of walking into the shop, so this is a *compound subject*.

Predicates

In addition to the subject, a sentence must also have a predicate. The *predicate* contains a verb and tells something about the subject. In addition to the verb, a predicate can also contain a direct or indirect object, object of a preposition, and other phrases.

> The cats napped on the front porch.

In this sentence, cats is the subject because the sentence is about cats.

The *complete predicate* is everything else in the sentence: *napped on the front porch*. This phrase is the predicate because it tells us what the cats did.

This sentence can be broken down into a simple subject and predicate:

> Cats napped.

In this sentence, *cats* is the simple subject, and *napped* is the *simple predicate*.

Although the sentence is very short and doesn't offer much information, it's still considered a complete sentence because it contains a subject and predicate.

Like a compound subject, a sentence can also have a **compound predicate**. This is when the subject is or does two or more things in the sentence.

> This easy chair reclines and swivels.

In this sentence, *this easy chair* is the complete subject. *Reclines and swivels* shows two actions of the chair, so this is the compound predicate.

Subject-Verb Agreement

The subject of a sentence and its verb must agree. The cornerstone rule of subject-verb agreement is that subject and verb must agree in number. Whether the subject is singular or plural, the verb must follow suit.

> Incorrect: The houses is new.
> Correct: The houses are new.
> Also Correct: The house is new.

In other words, a singular subject requires a singular verb; a plural subject requires a plural verb. The words or phrases that come between the subject and verb do not alter this rule.

> Incorrect: The houses built of brick is new.
> Correct: The houses built of brick are new.

> Incorrect: The houses with the sturdy porches is new.
> Correct: The houses with the sturdy porches are new.

The subject will always follow the verb when a sentence begins with *here* or *there*. Identify these with care.

Incorrect: Here *is* the *houses* with sturdy porches.
Correct: Here *are* the *houses* with sturdy porches.

The subject in the sentences above is not *here*, it is *houses*. Remember, *here* and *there* are never subjects. Be careful that contractions such as *here's* or *there're* do not cause confusion!

Two subjects joined by *and* require a plural verb form, except when the two combine to make one thing:

Incorrect: Garrett and Jonathan is over there.
Correct: Garrett and Jonathan are over there.

Incorrect: Spaghetti and meatballs are a delicious meal!
Correct: Spaghetti and meatballs is a delicious meal!

In the example above, *spaghetti and meatballs* is a compound noun. However, *Garrett and Jonathan* is not a compound noun.

Two singular subjects joined by *or, either/or,* or *neither/nor* call for a singular verb form.

Incorrect: Butter or syrup are acceptable.
Correct: Butter or syrup is acceptable.

Plural subjects joined by *or, either/or,* or *neither/nor* are, indeed, plural.

The chairs or the boxes are being moved next.

If one subject is singular and the other is plural, the verb should agree with the closest noun.

Correct: The chair or the boxes are being moved next.
Correct: The chairs or the box is being moved next.

Some plurals of money, distance, and time call for a singular verb.

Incorrect: Three dollars *are* enough to buy that.
Correct: Three dollars *is* enough to buy that.

For words declaring degrees of quantity such as *many of, some of,* or *most of,* let the noun that follows *of* be the guide:

Incorrect: Many of the books is in the shelf.
Correct: Many of the books are in the shelf.

Incorrect: Most of the pie *are* on the table.
Correct: Most of the pie *is* on the table.

For indefinite pronouns like anybody or everybody, use singular verbs.

Everybody *is* going to the store.

However, the pronouns *few, many, several, all, some,* and *both* have their own rules and use plural forms.

> Some *are* ready.

Some nouns like *crowd* and *congress* are called *collective nouns* and they require a singular verb form.

> Congress *is* in session.
> The news *is* over.

Books and movie titles, though, including plural nouns such as *Great Expectations*, also require a singular verb. Remember that only the subject affects the verb. While writing tricky subject-verb arrangements, say them aloud. Listen to them. Once the rules have been learned, one's ear will become sensitive to them, making it easier to pick out what's right and what's wrong.

Direct Objects

The *direct object* is the part of the sentence that receives the action of the verb. It is a noun and can usually be found after the verb. To find the direct object, first find the verb, and then ask the question *who* or *what* after it.

> The bear climbed the tree.

> What did the bear climb? *the tree*

Indirect Objects

An *indirect object* receives the direct object. It is usually found between the verb and the direct object. A strategy for identifying the indirect object is to find the verb and ask the questions *to whom/for whom* or *to what/ for what*.

> Jane made her daughter a cake.

> For whom did Jane make the cake? *her daughter*

Cake is the direct object because it is what Jane made, and *daughter* is the indirect object because she receives the cake.

Complements

A *complement* completes the meaning of an expression. A complement can be a pronoun, noun, or adjective. A verb complement refers to the direct object or indirect object in the sentence. An object complement gives more information about the direct object:

> The magician got the kids excited.

Kids is the direct object, and *excited* is the object complement.

A *subject complement* comes after a linking verb. It is typically an adjective or noun that gives more information about the subject:

> The king was noble and spared the thief's life.

Noble describes the *king* and follows the linking verb *was*.

Predicate Nouns
A *predicate noun* renames the subject:

John is a carpenter.

The subject is *John*, and the predicate noun is *carpenter*.

Predicate Adjectives
A *predicate adjective* describes the subject:

Margaret is beautiful.

The subject is *Margaret*, and the predicate adjective is *beautiful*.

Homonyms
Homonyms are words that sound the same but are spelled differently, and they have different meanings. There are several common homonyms that give writers trouble.

There, They're, and Their
The word *there* can be used as an adverb, adjective, or pronoun:

There are ten children on the swim team this summer.

I put my book over *there*, but now I can't find it.

The word *they're* is a contraction of the words *they* and *are*:

They're flying in from Texas on Tuesday.

The word *their* is a possessive pronoun:

I store *their* winter clothes in the attic.

Its and It's
Its is a possessive pronoun:

The cat licked *its* injured paw.

It's is the contraction for the words *it* and *is*:

It's unbelievable how many people opted not to vote in the last election.

Your and You're
Your is a possessive pronoun:

Can I borrow *your* lawnmower this weekend?

You're is a contraction for the words *you* and *are*:

You're about to embark on a fantastic journey.

To, Too, and Two

To is an adverb or a preposition used to show direction, relationship, or purpose:

> We are going *to* New York.

> They are going *to* see a show.

Too is an adverb that means more than enough, also, and very:

> You have had *too* much candy.

> We are on vacation that week, *too*.

Two is the written-out form of the numeral 2:

> *Two* of the shirts didn't fit, so I will have to return them.

New and Knew

New is an adjective that means recent:

> There's a *new* customer on the phone.

Knew is the past tense of the verb *know*:

> I *knew* you'd have fun on this ride.

Affect and Effect

Affect and *effect* are complicated because they are used as both nouns and verbs, have similar meanings, and are pronounced the same.

	Affect	**Effect**
Noun Definition	emotional state	result
Noun Example	The patient's affect was flat.	The effects of smoking are well documented.
Verb Definition	to influence	to bring about
Verb Example	The pollen count affects my allergies.	The new candidate hopes to effect change.

Independent and Dependent Clauses

Independent and *dependent* clauses are strings of words that contain both a subject and a verb. An independent clause *can* stand alone as complete thought, but a dependent clause *cannot*. A dependent clause relies on other words to be a complete sentence.

> Independent clause: The keys are on the counter.
> Dependent clause: If the keys are on the counter

Notice that both clauses have a subject (*keys*) and a verb (*are*). The independent clause expresses a complete thought, but the word *if* at the beginning of the dependent clause makes it *dependent* on other words to be a complete thought.

> Independent clause: If the keys are on the counter, please give them to me.

This presents a complete sentence since it includes at least one verb and one subject and is a complete thought. In this case, the independent clause has two subjects (*keys* & an implied *you*) and two verbs (*are* & *give*).

> Independent clause: I went to the store.
> Dependent clause: Because we are out of milk,
>
> Complete Sentence: Because we are out of milk, I went to the store.
> Complete Sentence: I went to the store because we are out of milk.

Phrases

A *phrase* is a group of words that do not make a complete thought or a clause. They are parts of sentences or clauses. Phrases can be used as nouns, adjectives, or adverbs. A phrase does not contain both a subject and a verb.

Prepositional Phrases

A *prepositional phrase* shows the relationship between a word in the sentence and the object of the preposition. The object of the preposition is a noun that follows the preposition.

> The orange pillows are on the couch.

On is the preposition, and *couch* is the object of the preposition.

> She brought her friend with the nice car.

With is the preposition, and *car* is the object of the preposition. Here are some common prepositions:

about	as	at	after
by	for	from	in
of	on	to	with

Verbals and Verbal Phrases

Verbals are forms of verbs that act as other parts of speech. They can be used as nouns, adjectives, or adverbs. Though they are use verb forms, they are not to be used as the verb in the sentence. A word group that is based on a verbal is considered a *verbal phrase*. There are three major types of verbals: *participles*, *gerunds*, and *infinitives*.

Participles are verbals that act as adjectives. The present participle ends in *–ing*, and the past participle ends in *–d, -ed, -n, or-t*.

Verb	Present Participle	Past Participle
walk	walking	walked
share	sharing	shared

Participial phrases are made up of the participle and modifiers, complements, or objects.

> Crying for most of an hour, the baby didn't seem to want to nap.

> Having already taken this course, the student was bored during class.

> *Crying for most of an hour* and *Having already taken this course* are the participial phrases.

Gerunds are verbals that are used as nouns and end in *–ing*. A gerund can be the subject or object of the sentence like a noun. Note that a present participle can also end in *–ing*, so it is important to distinguish between the two. The gerund is used as a noun, while the participle is used as an adjective.

Swimming is my favorite sport.

I wish I were sleeping.

A *gerund phrase* includes the gerund and any modifiers or complements, direct objects, indirect objects, or pronouns.

Cleaning the house is my least favorite weekend activity.

Cleaning the house is the gerund phrase acting as the subject of the sentence.

The most important goal this year is raising money for charity.

Raising money for charity is the gerund phrase acting as the direct object.

The police accused the woman of stealing the car.

The *gerund* phrase *stealing the car* is the object of the preposition in this sentence.

An *infinitive* is a verbal made up of the word to and a verb. Infinitives can be used as nouns, adjectives, or adverbs.

Examples: To eat, to jump, to swim, to lie, to call, to work

An *infinitive phrase* is made up of the infinitive plus any complements or modifiers. The infinitive phrase *to wait* is used as the subject in this sentence:

To wait was not what I had in mind.

The infinitive phrase *to sing* is used as the subject complement in this sentence:

Her dream is to sing.

The infinitive phrase *to grow* is used as an adverb in this sentence:

Children must eat to grow.

Appositive Phrases
An *appositive* is a noun or noun phrase that renames a noun that comes immediately before it in the sentence. An appositive can be a single word or several words. These phrases can be *essential* or *nonessential*. An essential appositive phrase is necessary to the meaning of the sentence and a nonessential appositive phrase is not. It is important to be able to distinguish these for purposes of comma use.

Essential: My sister Christina works at a school.

Naming which sister is essential to the meaning of the sentence, so no commas are needed.

Nonessential: My sister, who is a teacher, is coming over for dinner tonight.

Who is a teacher is not essential to the meaning of the sentence, so commas are required.

Absolute Phrases

An *absolute phrase* modifies a noun without using a conjunction. It is not the subject of the sentence and is not a complete thought on its own. Absolute phrases are set off from the independent clause with a comma.

> *Arms outstretched,* she yelled at the sky.

> *All things considered*, this has been a great day.

Demonstrate Command of the Conventions of Standard English Capitalization, Punctuation, and Spelling When Writing

Capitalization

Here's a non-exhaustive list of things that should be capitalized.

- The first word of every sentence
- The first word of every line of poetry
- The first letter of proper nouns (World War II)
- Holidays (Valentine's Day)
- The days of the week and months of the year (Tuesday, March)
- The first word, last word, and all major words in the titles of books, movies, songs, and other creative works (In the novel, *To Kill a Mockingbird*, note that *a* is lowercase since it's not a major word, but *to* is capitalized since it's the first word of the title.)
- Titles when preceding a proper noun (President Roberto Gonzales, Aunt Judy)

When simply using a word such as president or secretary, though, the word is not capitalized.

> Officers of the new business must include a *president* and *treasurer*.

Seasons—spring, fall, etc.—are not capitalized.

North, south, east, and *west* are capitalized when referring to regions but are not when being used for directions. In general, if it's preceded by *the* it should be capitalized.

> I'm from the South.
> I drove south.

Punctuation

Ellipses

An *ellipsis* (. . .) consists of three handy little dots that can speak volumes on behalf of irrelevant material. Writers use them in place of words, lines, phrases, list content, or paragraphs that might just as easily have been omitted from a passage of writing. This can be done to save space or to focus only on the specifically relevant material.

> Exercise is good for some unexpected reasons. Watkins writes, "Exercise has many benefits such as . . . reducing cancer risk."

In the example above, the ellipsis takes the place of the other benefits of exercise that are more expected.

The ellipsis may also be used to show a pause in sentence flow.

"I'm wondering . . . how this could happen," Dylan said in a soft voice.

Commas

A *comma* (,) is the punctuation mark that signifies a pause—breath—between parts of a sentence. It denotes a break of flow. As with so many aspects of writing structure, authors will benefit by reading their writing aloud or mouthing the words. This can be particularly helpful if one is uncertain about whether the comma is needed.

In a complex sentence—one that contains a subordinate (dependent) clause or clauses—the use of a comma is dictated by where the subordinate clause is located. If the subordinate clause is located before the main clause, a comma is needed between the two clauses.

I will not pay for the steak, *because I don't have that much money.*

Generally, if the subordinate clause is placed after the main clause, no punctuation is needed. I did well on my exam because I studied two hours the night before. Notice how the last clause is dependent, because it requires the earlier independent clauses to make sense.

Use a comma on both sides of an interrupting phrase.

I will pay for the ice cream, chocolate and vanilla, and then will eat it all myself.

The words forming the phrase in italics are nonessential (extra) information. To determine if a phrase is nonessential, try reading the sentence without the phrase and see if it's still coherent.

A comma is not necessary in this next sentence because no interruption—nonessential or extra information—has occurred. Read sentences aloud when uncertain.

I will pay for his chocolate and vanilla ice cream and then will eat it all myself.

If the nonessential phrase comes at the beginning of a sentence, a comma should only go at the end of the phrase. If the phrase comes at the end of a sentence, a comma should only go at the beginning of the phrase.

Other types of interruptions include the following:

- interjections: Oh no, I am not going.
- abbreviations: Barry Potter, M.D., specializes in heart disorders.
- direct addresses: Yes, Claudia, I am tired and going to bed.
- parenthetical phrases: His wife, lovely as she was, was not helpful.
- transitional phrases: Also, it is not possible.

The second comma in the following sentence is called an Oxford comma.

I will pay for ice cream, syrup, and pop.

It is a comma used after the second-to-last item in a series of three or more items. It comes before the word *or* or *and*. Not everyone uses the Oxford comma; it is optional, but many believe it is needed. The comma functions as a tool to reduce confusion in writing. So, if omitting the Oxford comma would cause confusion, then it's best to include it.

Commas are used in math to mark the place of thousands in numerals, breaking them up so they are easier to read. Other uses for commas are in dates (*March 19, 2016*), letter greetings (*Dear Sally,*), and in between cities and states (*Louisville, KY*).

Semicolons

The *semicolon* (;) might be described as a heavy-handed comma. Take a look at these two examples:

> I will pay for the ice cream, but I will not pay for the steak.
> I will pay for the ice cream; I will not pay for the steak.

What's the difference? The first example has a comma and a conjunction separating the two independent clauses. The second example does not have a conjunction, but there are two independent clauses in the sentence. So something more than a comma is required. In this case, a semicolon is used.

Two independent clauses can only be joined in a sentence by either a comma and conjunction or a semicolon. If one of those tools is not used, the sentence will be a run-on. Remember that while the clauses are independent, they need to be closely related in order to be contained in one sentence.

Another use for the semicolon is to separate items in a list when the items themselves require commas.

> The family lived in Phoenix, Arizona; Oklahoma City, Oklahoma; and Raleigh, North Carolina.

Colons

Colons have many miscellaneous functions. Colons can be used to precede further information or a list. In these cases, a colon should only follow an independent clause.

> Humans take in sensory information through five basic senses: sight, hearing, smell, touch, and taste.

> The meal includes the following components:

- Caesar salad
- spaghetti
- garlic bread
- cake

> The family got what they needed: a reliable vehicle.

While a comma is more common, a colon can also precede a formal quotation.

> He said to the crowd: "Let's begin!"

The colon is used after the greeting in a formal letter.

> Dear Sir:
> To Whom It May Concern:

In the writing of time, the colon separates the minutes from the hour (*4:45 p.m.*). The colon can also be used to indicate a ratio between two numbers (*50:1*).

Hyphens

The *hyphen* (-) is a little hash mark that can be used to join words to show that they are linked.

Hyphenate two words that work together as a single adjective (a compound adjective).

>   honey-covered biscuits

Some words always require hyphens, even if not serving as an adjective.

>   merry-go-round

Hyphens always go after certain prefixes like *anti-* & *all-*.

Hyphens should also be used when the absence of the hyphen would cause a strange vowel combination (*semi-engineer*) or confusion. For example, *re-collect* should be used to describe something being gathered twice rather than being written as *recollect*, which means to remember.

Parentheses and Dashes

Parentheses are half-round brackets that look like this: ( ). They set off a word, phrase, or sentence that is a afterthought, explanation, or side note relevant to the surrounding text but not essential. A pair of commas is often used to set off this sort of information, but parentheses are generally used for information that would not fit well within a sentence or that the writer deems not important enough to be structurally part of the sentence.

>   The picture of the heart (see above) shows the major parts you should memorize.
>   Mount Everest is one of three mountains in the world that are over 28,000 feet high (K2 and Kanchenjunga are the other two).

See how the sentences above are complete without the parenthetical statements? In the first example, *see above* would not have fit well within the flow of the sentence. The second parenthetical statement could have been a separate sentence, but the writer deemed the information not pertinent to the topic.

The dash (—) is a mark longer than a hyphen used as a punctuation mark in sentences and to set apart a relevant thought. Even after plucking out the line separated by the dash marks, the sentence will be intact and make sense.

>   Looking out the airplane window at the landmarks—Lake Clarke, Thompson Community College, and the bridge—she couldn't help but feel excited to be home.

The dashes use is similar to that of parentheses or a pair of commas. So, what's the difference? Many believe that using dashes makes the clause within them stand out while using parentheses is subtler. It's advised to not use dashes when commas could be used instead.

Quotation Marks

Here are some instances where *quotation marks* should be used:

- Dialogue for characters in narratives. When characters speak, the first word should always be capitalized and the punctuation goes inside the quotes. For example:

  Janie said, "The tree fell on my car during the hurricane."

- Around titles of songs, short stories, essays, and chapter in books
- To emphasize a certain word
- To refer to a word as the word itself

Apostrophes

This punctuation mark, the apostrophe ('), is a versatile little mark. It has a few different functions:

- Quotes: Apostrophes are used when a second quote is needed within a quote.

  In my letter to my friend, I wrote, "The girl had to get a new purse, and guess what Mary did? She said, 'I'd like to go with you to the store.' I knew Mary would buy it for her."

- Contractions: Another use for an apostrophe in the quote above is a contraction. *I'd* is used for *I would*.

- Possession: An apostrophe followed by the letter *s* shows possession (*Mary's* purse). If the possessive word is plural, the apostrophe generally just follows the word.

  The trees' leaves are all over the ground.

Spelling

Spelling might or might not be important to some, or maybe it just doesn't come naturally, but those who are willing to discover some new ideas and consider their benefits can learn to spell better and improve their writing. Misspellings reduce a writer's credibility and can create misunderstandings. Spell checkers built into word processors are not a substitute for accuracy. They are neither foolproof nor without error. In addition, a writer's misspelling of one word may also be a word. For example, a writer intending to spell *herd* might accidentally type *s* instead of *d* and unintentionally spell *hers*. Since *hers* is a word, it would not be marked as a misspelling by a spell checker. In short, use spell check, but don't rely on it.

Guidelines for Spelling

Saying and listening to a word serves as the beginning of knowing how to spell it. Keep these subsequent guidelines in mind, remembering there are often exceptions because the English language is replete with them.

Guideline #1: Syllables must have at least one vowel. In fact, every syllable in every English word has a vowel.

- dog
- haystack
- answering
- abstentious
- simple

Guideline #2: The long and short of it. When the vowel has a short vowel sound as in *mad* or *bed,* only the single vowel is needed. If the word has a long vowel sound, add another vowel, either alongside it or separated by a consonant: bed/*bead*; mad/*made.* When the second vowel is separated by two spaces— *madder*—it does not affect the first vowel's sound.

Guideline #3: Suffixes. Refer to the examples listed above.

Guideline #4: Which comes first; the *i* or the *e*? Remember the saying, "*I* before *e* except after *c* or when sounding as *a* as in *neighbor* or *weigh.*" Keep in mind that these are only guidelines and that there are always exceptions to every rule.

Guideline #5: Vowels in the right order. Another helpful rhyme is, "When two vowels go walking, the first one does the talking." When two vowels are in a row, the first one often has a long vowel sound and the other is silent. An example is *team.*

If you have difficulty spelling words, determine a strategy to help. Work on spelling by playing word games like Scrabble or Words with Friends. Consider using phonics, which is sounding words out by slowly and surely stating each syllable. Try repeating and memorizing spellings as well as picturing words in your head. Try making up silly memory aids. See what works best.

Homophones

Homophones are two or more words that have no particular relationship to one another except their identical pronunciations. Homophones make spelling English words fun and challenging like these:

Common Homophones		
affect, effect	cell, sell	it's, its
allot, a lot	do, due, dew	knew, new
barbecue, barbeque	dual, duel	libel, liable
bite, byte	eminent, imminent	principal, principle
brake, break	flew, flu, flue	their, there, they're
capital, capitol	gauge, gage	to, too, two
cash, cache	holy, wholly	yoke, yolk

Irregular Plurals

Irregular plurals are words that aren't made plural the usual way.

- Most nouns are made plural by adding –*s* (book*s*, television*s*, skyscraper*s*).

- Most nouns ending in *ch, sh, s, x,* or *z* are made plural by adding –*es* (church*es*, marsh*es*).

- Most nouns ending in a vowel + *y* are made plural by adding –*s* (day*s*, toy*s*).

- Most nouns ending in a consonant + *y*, are made plural by the -*y* becoming -*ies* (baby becomes *babies*).

- Most nouns ending in an *o* are made plural by adding –*s* (piano*s*, photo*s*).

- Some nouns ending in an *o*, though, may be made plural by adding –*es* (example: potato*es*, volcano*es*), and, of note, there is no known rhyme or reason for this!

- Most nouns ending in an *f* or *fe* are made plural by the -*f* or -*fe* becoming -*ves*! (example: wolf becomes *wolves*).

- Some words function as both the singular and plural form of the word (fish, deer).

- Other exceptions include *man* becomes *men, mouse* becomes *mice, goose* becomes *geese,* and *foot* becomes *feet.*

Contractions

The basic rule for making *contractions* is one area of spelling that is pretty straightforward: combine the two words by inserting an apostrophe (') in the space where a letter is omitted. For example, to combine *you* and *are*, drop the *a* and put the apostrophe in its place: *you're*.

> he + is = he's
> you + all = y'all (informal but often misspelled)

Note that *it's*, when spelled with an apostrophe, is always the contraction for *it is*. The possessive form of the word is written without an apostrophe as *its*.

Correcting Misspelled Words

A good place to start looking at commonly misspelled words here is with the word *misspelled*. While it looks peculiar, look at it this way: *mis* (the prefix meaning *wrongly*) + *spelled* = *misspelled*.

Let's look at some commonly misspelled words.

Commonly Misspelled Words

accept	benign	existence	jewelry	parallel	separate
acceptable	bicycle	experience	judgment	pastime	sergeant
accidentally	brief	extraordinary	library	permissible	similar
accommodate	business	familiar	license	perseverance	supersede
accompany	calendar	February	maintenance	personnel	surprise
acknowledgement	campaign	fiery	maneuver	persuade	symmetry
acquaintance	candidate	finally	mathematics	possess	temperature
acquire	category	forehead	mattress	precede	tragedy
address	cemetery	foreign	millennium	prevalent	transferred
aesthetic	changeable	foremost	miniature	privilege	truly
aisle	committee	forfeit	mischievous	pronunciation	usage
altogether	conceive	glamorous	misspell	protein	valuable
amateur	congratulations	government	mortgage	publicly	vengeance
apparent	courtesy	grateful	necessary	questionnaire	villain
appropriate	deceive	handkerchief	neither	recede	Wednesday
arctic	desperate	harass	nickel	receive	weird
asphalt	discipline	hygiene	niece	recommend	
associate	disappoint	hypocrisy	ninety	referral	
attendance	dissatisfied	ignorance	noticeable	relevant	
auxiliary	eligible	incredible	obedience	restaurant	
available	embarrass	intelligence	occasion	rhetoric	
balloon	especially	intercede	occurrence	rhythm	
believe	exaggerate	interest	omitted	schedule	
beneficial	exceed	irresistible	operate	sentence	

Practice Questions

1. What is the structure of the following sentence?
   The restaurant is unconventional because it serves both Chicago style pizza and New York style pizza.

   a. Simple
   b. Compound
   c. Complex
   d. Compound-complex

2. The following sentence contains what kind of error?
   This summer, I'm planning to travel to Italy, take a Mediterranean cruise, going to Pompeii, and eat a lot of Italian food.

   a. Parallelism
   b. Sentence fragment
   c. Misplaced modifier
   d. Subject-verb agreement

3. The following sentence contains what kind of error?
   Forgetting that he was supposed to meet his girlfriend for dinner, Anita was mad when Fred showed up late.

   a. Parallelism
   b. Run-on sentence
   c. Misplaced modifier
   d. Subject-verb agreement

4. The following sentence contains what kind of error?
   Some workers use all their sick leave, other workers cash out their leave.

   a. Parallelism
   b. Comma splice
   c. Sentence fragment
   d. Subject-verb agreement

5. A student writes the following in an essay:

Protestors filled the streets of the city. Because they were dissatisfied with the government's leadership.

Which of the following is an appropriately-punctuated correction for this sentence?

a. Protestors filled the streets of the city, because they were dissatisfied with the government's leadership.

b. Protesters, filled the streets of the city, because they were dissatisfied with the government's leadership.

c. Because they were dissatisfied with the government's leadership protestors filled the streets of the city.

d. Protestors filled the streets of the city because they were dissatisfied with the government's leadership.

6. What is the part of speech of the underlined word in the sentence?

We need to come up with a fresh <u>approach</u> to this problem.

a. Noun
b. Verb
c. Adverb
d. Adjective

7. What is the part of speech of the underlined word in the sentence?

Investigators conducted an <u>exhaustive</u> inquiry into the accusations of corruption.

a. Noun
b. Verb
c. Adverb
d. Adjective

8. The underlined portion of the sentence is an example of which sentence component?

New students should report <u>to the student center</u>.

a. Dependent clause
b. Adverbial phrase
c. Adjective clause
d. Noun phrase

9. What is the noun phrase in the following sentence?

Charlotte's new German shepherd puppy is energetic.

a. Puppy
b. Charlotte
c. German shepherd puppy
d. Charlotte's new German shepherd puppy

10. Which word choices will correctly complete the sentence?

   Increasing the price of bus fares has had a greater [affect / effect] on ridership [then / than] expected.

   a. affect; then
   b. affect; than
   c. effect; then
   d. effect; than

11. While studying vocabulary, a student notices that the words *circumference*, *circumnavigate*, and *circumstance* all begin with the prefix *circum–*. The student uses her knowledge of affixes to infer that all of these words share what related meaning?
   a. Around, surrounding
   b. Travel, transport
   c. Size, measurement
   d. Area, location

12. A student wants to rewrite the following sentence:

   Entrepreneurs use their ideas to make money.

He wants to use the word *money* as a verb, but he isn't sure which word ending to use. What is the appropriate suffix to add to *money* to complete the following sentence?

   Entrepreneurs _____ their ideas.

   a. –ize
   b. –ical
   c. –en
   d. –ful

13. A student reads the following sentence:

   A hundred years ago, automobiles were rare, but now cars are ubiquitous.

However, she doesn't know what the word *ubiquitous* means. Which key context clue is essential to decipher the word's meaning?

   a. Ago
   b. Cars
   c. Now
   d. Rare

14. A local newspaper is looking for writers for a student column. A student would like to submit his article to the newspaper, but he isn't sure how to format his article according to journalistic standards. What resource should he use?
   a. A thesaurus
   b. A dictionary
   c. A style guide
   d. A grammar book

15. A student encounters the word *aficionado* and wants to learn more about it. It doesn't sound like other English words he knows, so the student is curious to identify the word's origin. What resource should he consult?
    a. A thesaurus
    b. A dictionary
    c. A style guide
    d. A grammar book

16. Which domain is likely to be used by a website run by a nonprofit group?
    a. .com
    b. .edu
    c. .org
    d. .gov

17. Which mode of writing aims to inform the reader objectively about a particular subject or idea and typically contains definitions, instructions, or facts within its subject matter?
    a. Argumentative
    b. Informative
    c. Narrative
    d. Descriptive

18. Editorials, letters of recommendation, and cover letters most likely incorporate which writing mode?
    a. Argumentative
    b. Informative
    c. Narrative
    d. Descriptive

19. The type of writing mode an author chooses to use is dependent on which of the following elements?
    a. The audience
    b. The primary purpose
    c. The main idea
    d. Both A and B

20. The rhetorical appeal that elicits an emotional and/or sympathetic response from an audience is known as which of the following?
    a. Logos
    b. Ethos
    c. Pathos
    d. None of the above

21. Which of the following refers to what an author wants to express about a given subject?
    a. Primary purpose
    b. Plot
    c. Main idea
    d. Characterization

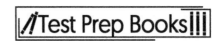

22. Which organizational style is used in the following passage?

     There are several reasons why the new student café has not been as successful as expected. One factor is that prices are higher than originally advertised, so many students cannot afford to buy food and beverages there. Also, the café closes rather early; as a result, students go out in town to other late-night gathering places rather than meeting friends at the café on campus.

    a. Cause and effect order
    b. Compare and contrast order
    c. Spatial order
    d. Time order

23. Short, succinct sentences are best written for which of the following audiences?
    a. Adults or people more familiar with a subject
    b. Children or people less familiar a subject
    c. Politicians and academics
    d. University students

24. A student is starting a research assignment on Japanese-American internment camps during World War II, but she is unsure of how to gather relevant resources. Which of the following would be the most helpful advice for the student?
    a. Conduct a broad internet search to get a wide view of the subject.
    b. Consult an American history textbook.
    c. Find websites about Japanese culture such as fashion and politics.
    d. Locate texts in the library related to World War II in America and look for references to internment camps in the index.

25. Which of the following should be evaluated to ensure the credibility of a source?
    a. The publisher, the author, and the references
    b. The subject, the title, and the audience
    c. The organization, stylistic choices, and transition words
    d. The length, the tone, and the contributions of multiple authors

26. Which of the following sentences has an error in capitalization?
    a. The East Coast has experienced very unpredictable weather this year.
    b. My Uncle owns a home in Florida, where he lives in the winter.
    c. I am taking English Composition II on campus this fall.
    d. There are several nice beaches we can visit on our trip to the Jersey Shore this summer.

27. Julia Robinson, an avid photographer in her spare time, was able to capture stunning shots of the local wildlife on her last business trip to Australia.
Which of the following is an adjective in the preceding sentence?
    a. Time
    b. Capture
    c. Avid
    d. Photographer

28. Which of the following sentences uses correct punctuation?
    a. Carole is not currently working; her focus is on her children at the moment.
    b. Carole is not currently working and her focus is on her children at the moment.
    c. Carole is not currently working, her focus is on her children at the moment.
    d. Carole is not currently working her focus is on her children at the moment.

29. Which of these examples is a compound sentence?
    a. Alex and Shane spent the morning coloring and later took a walk down to the park.
    b. After coloring all morning, Alex and Shane spent the afternoon at the park.
    c. Alex and Shane spent the morning coloring, and then they took a walk down to the park.
    d. After coloring all morning and spending part of the day at the park, Alex and Shane took a nap.

30. Which of these examples shows incorrect use of subject-verb agreement?
    a. Neither of the cars are parked on the street.
    b. Both of my kids are going to camp this summer.
    c. Any of your friends are welcome to join us on the trip in November.
    d. Each of the clothing options is appropriate for the job interview.

31. When it gets warm in the spring, _____ and _____ like to go fishing at Cobbs Creek.
Which of the following word pairs should be used in the blanks above?
    a. me, him
    b. he, I
    c. him, I
    d. he, me

32. Which example shows correct comma usage for dates?
    a. The due date for the final paper in the course is Monday, May 16, 2016.
    b. The due date for the final paper in the course is Monday, May 16 2016.
    c. The due date for the final project in the course is Monday, May, 16, 2016.
    d. The due date for the final project in the course is Monday May 16, 2016.

33. Which of the following uses correct spelling?
    a. Leslie knew that training for the Philadelphia Marathon would take dicsipline and perserverance, but she was up to the challenge.
    b. Leslie knew that training for the Philadelphia Marathon would take discipline and perseverence, but she was up to the challenge.
    c. Leslie knew that training for the Philadelphia Marathon would take disiplin and perservearance, but she was up to the challenge.
    d. Leslie knew that training for the Philadelphia Marathon would take discipline and perseverance, but she was up to the challenge.

34. At last night's company function, in honor of Mr. Robertson's retirement, several employees spoke kindly about his career achievements.

In the preceding sentence, what part of speech is the word *function*?
    a. Adjective
    b. Adverb
    c. Verb
    d. Noun

35. Which of the examples uses the correct plural form?
    a. Tomatos
    b. Analysis
    c. Cacti
    d. Criterion

36. Which of the following examples uses correct punctuation?
    a. The moderator asked the candidates, "Is each of you prepared to discuss your position on global warming?".
    b. The moderator asked the candidates, "Is each of you prepared to discuss your position on global warming?"
    c. The moderator asked the candidates, 'Is each of you prepared to discuss your position on global warming?'
    d. The moderator asked the candidates, "Is each of you prepared to discuss your position on global warming"?

37. Based on the words *transfer, transact, translation, transport*, what is the meaning of the prefix *trans*?
    a. Separation
    b. All, everywhere
    c. Forward
    d. Across, beyond, over

38. In which of the following sentences does the word *part* function as an adjective?
    a. The part Brian was asked to play required many hours of research.
    b. She parts ways with the woodsman at the end of the book.
    c. The entire team played a part in the success of the project.
    d. Ronaldo is part Irish on his mother's side of the family.

39. All of Shannon's family and friends helped her to celebrate her 50th birthday at Café Sorrento. Which of the following is the complete subject of the preceding sentence?
    a. Family and friends
    b. All
    c. All of Shannon's family and friends
    d. Shannon's family and friends

40. Which of the following sentences uses second person point of view?
    a. I don't want to make plans for the weekend before I see my work schedule.
    b. She had to miss the last three yoga classes due to illness.
    c. Pluto is no longer considered a planet because it is not gravitationally dominant.
    d. Be sure to turn off all of the lights before locking up for the night.

41. As the tour group approached the bottom of Chichen Itza, the prodigious Mayan pyramid, they became nervous about climbing its distant peak.

Based on the context of the sentence, which of the following words shows the correct meaning of the word *prodigious*?
    a. Very large
    b. Famous
    c. Very old
    d. Fancy

42. Which of the following sentences correctly uses a hyphen?
    a. Last-year, many of the players felt unsure of the coach's methods.
    b. Some of the furniture she selected seemed a bit over - the - top for the space.
    c. Henry is a beagle-mix and is ready for adoption this weekend.
    d. Geena works to maintain a good relationship with her ex-husband to the benefit of their children.

43. Which of the following examples correctly uses quotation marks?
    a. "Where the Red Fern Grows" was one of my favorite novels as a child.
    b. Though he is famous for his roles in films like "The Great Gatsby" and "Titanic," Leonardo DiCaprio has never won an Oscar.
    c. Sylvia Plath's poem, "Daddy" will be the subject of this week's group discussion.
    d. "The New York Times" reported that many fans are disappointed in some of the trades made by the Yankees this off-season.

44. Which of the following sentences shows correct word usage?
    a. It's often been said that work is better then rest.
    b. Its often been said that work is better then rest.
    c. It's often been said that work is better than rest.
    d. Its often been said that work is better than rest.

45. Glorify, fortify, gentrify, acidify

Based on the preceding words, what is the correct meaning of the suffix –*fy*?
    a. Marked by, given to
    b. Doer, believer
    c. Make, cause, cause to have
    d. Process, state, rank

46. Which of the following uses correct spelling?
    a. Jed was disatified with the acommodations at his hotel, so he requested another room.
    b. Jed was dissatisfied with the accommodations at his hotel, so he requested another room.
    c. Jed was dissatisfied with the accomodations at his hotel, so he requested another room.
    d. Jed was disatified with the accommodations at his hotel, so he requested another room.

47. Which of the following is an imperative sentence?
    a. Pennsylvania's state flag includes two draft horses and an eagle.
    b. Go down to the basement and check the hot water heater for signs of a leak.
    c. You must be so excited to have a new baby on the way!
    d. How many countries speak Spanish?

48. After a long day at work, Tracy had dinner with her family, and then took a walk to the park.
What are the transitional words in the preceding sentence?
   a. After, then
   b. At, with, to
   c. Had, took
   d. A, the

49. Which of the following examples is a compound sentence?
   a. Shawn and Jerome played soccer in the backyard for two hours.
   b. Marissa last saw Elena and talked to her this morning.
   c. The baby was sick, so I decided to stay home from work.
   d. Denise, Kurt, and Eric went for a run after dinner.

50. Robert needed to find at least four sources for his final project, so he searched several library databases for reliable academic research.
Which words function as nouns in the preceding sentence?
   a. Robert, sources, project, databases, research
   b. Robert, sources, final, project, databases, academic, research
   c. Robert, sources, project, he, library, databases, research
   d. Sources, project, databases, research

51. Which of the following sentences uses correct subject-verb agreement?
   a. There is two constellations that can be seen from the back of the house.
   b. At least four of the sheep needs to be sheared before the end of summer.
   c. Lots of people were auditioning for the singing competition on Saturday.
   d. Everyone in the group have completed the assignment on time.

52. Philadelphia is home to some excellent walking tours where visitors can learn more about the culture and rich history of the city of brotherly love.
What are the adjectives in the preceding sentence?
   a. Philadelphia, tours, visitors, culture, history, city, love
   b. Excellent, walking, rich, brotherly
   c. Is, can, learn
   d. To, about, of

53. The realtor showed _____ and _____ a house on Wednesday afternoon.
Which of the following pronoun pairs should be used in the blanks above?
   a. She, I
   b. She, me
   c. Me, her
   d. Her, me

54. Which of the following examples uses correct punctuation?
   a. Recommended supplies for the hunting trip include the following: rain gear, large backpack, hiking boots, flashlight, and non-perishable foods.
   b. I left the store, because I forgot my wallet.
   c. As soon as the team checked into the hotel; they met in the lobby for a group photo.
   d. None of the furniture came in on time: so they weren't able to move in to the new apartment.

55. Which of the following sentences shows correct word usage?
    a. Your going to have to put you're jacket over their.
    b. You're going to have to put your jacket over there.
    c. Your going to have to put you're jacket over they're.
    d. You're going to have to put your jacket over their.

Answer Explanations

1. C: A complex sentence joins an independent or main clause with a dependent or subordinate clause. In this case, the main clause is "The restaurant is unconventional." This is a clause with one subject-verb combination that can stand alone as a grammatically-complete sentence. The dependent clause is "because it serves both Chicago style pizza and New York style pizza." This clause begins with the subordinating conjunction *because* and also consists of only one subject-verb combination. *A* is incorrect because a simple sentence consists of only one verb-subject combination—one independent clause. *B* is incorrect because a compound sentence contains two independent clauses connected by a conjunction. *D* is incorrect because a complex-compound sentence consists of two or more independent clauses and one or more dependent clauses.

2. A: Parallelism refers to consistent use of sentence structure or word form. In this case, the list within the sentence does not utilize parallelism; three of the verbs appear in their base form—*travel*, *take*, and *eat*—but one appears as a gerund—*going*. A parallel version of this sentence would be "This summer, I'm planning to travel to Italy, take a Mediterranean cruise, go to Pompeii, and eat a lot of Italian food." *B* is incorrect because this description is a complete sentence. *C* is incorrect as a misplaced modifier is a modifier that is not located appropriately in relation to the word or words they modify. *D* is incorrect because subject-verb agreement refers to the appropriate conjugation of a verb in relation to its subject.

3. C: In this sentence, the modifier is the phrase "Forgetting that he was supposed to meet his girlfriend for dinner." This phrase offers information about Fred's actions, but the noun that immediately follows it is Anita, creating some confusion about the "do-er" of the phrase. A more appropriate sentence arrangement would be "Forgetting that he was supposed to meet his girlfriend for dinner, Fred made Anita mad when he showed up late." *A* is incorrect as parallelism refers to the consistent use of sentence structure and verb tense, and this sentence is appropriately consistent. *B* is incorrect as a run-on sentence does not contain appropriate punctuation for the number of independent clauses presented, which is not true of this description. *D* is incorrect because subject-verb agreement refers to the appropriate conjugation of a verb relative to the subject, and all verbs have been properly conjugated.

4. B: A comma splice occurs when a comma is used to join two independent clauses together without the additional use of an appropriate conjunction. One way to remedy this problem is to replace the comma with a semicolon. Another solution is to add a conjunction: "Some workers use all their sick leave, but other workers cash out their leave." *A* is incorrect as parallelism refers to the consistent use of sentence structure and verb tense; all tenses and structures in this sentence are consistent. *C* is incorrect because a sentence fragment is a phrase or clause that cannot stand alone—this sentence contains two independent clauses. *D* is incorrect because subject-verb agreement refers to the proper conjugation of a verb relative to the subject, and all verbs have been properly conjugated.

5. D: The problem in the original passage is that the second sentence is a dependent clause that cannot stand alone as a sentence; it must be attached to the main clause found in the first sentence. Because the main clause comes first, it does not need to be separated by a comma. However, if the dependent clause came first, then a comma would be necessary, which is why Choice *C* is incorrect. *A* and *B* also insert unnecessary commas into the sentence.

6. A: A noun refers to a person, place, thing, or idea. Although the word *approach* can also be used as a verb, in the sentence it functions as a noun within the noun phrase "a fresh approach," so *B* is incorrect. An adverb is a word or phrase that provides additional information of the verb, but because the verb is *need* and not *approach*, then *C* is false. An adjective is a word that describes a noun, used here as the word *fresh*, but it is not the noun itself. Thus, *D* is also incorrect.

7. D: An adjective modifies a noun, answering the question "Which one?" or "What kind?" In this sentence, the word *exhaustive* is an adjective that modifies the noun *investigation*. Another clue that this word is an adjective is the suffix *–ive*, which means "having the quality of." The nouns in this sentence are investigators, inquiry, accusations, and corruption; therefore, *A* is incorrect. The verb in this sentence is *conducted* because this was the action taken by the subject *the investigators*; therefore, *B* is incorrect. *C* is incorrect because an adverb is a word or phrase that provides additional information about the verb, expressing how, when, where, or in what manner.

8. B: In this case, the phrase functions as an adverb modifying the verb *report*, so *B* is the correct answer. "To the student center" does not consist of a subject-verb combination, so it is not a clause; thus, Choices *A* and *C* can be eliminated. This group of words is a phrase. Phrases are classified by either the controlling word in the phrase or its function in the sentence. *D* is incorrect because a noun phrase is a series of words that describe or modify a noun.

9. D: A noun phrase consists of the noun and all of its modifiers. In this case, the subject of the sentence is the noun *puppy*, but it is preceded by several modifiers—adjectives that give more information about what kind of puppy, which are also part of the noun phrase. Thus, *A* is incorrect. Charlotte is the owner of the puppy and a modifier of the puppy, so *B* is false. *C* is incorrect because it contains some, but not all, of the modifiers pertaining to the puppy. *D* is correct because it contains all of them.

10. D: In this sentence, the first answer choice requires a noun meaning *impact* or *influence*, so *effect* is the correct answer. For the second answer choice, the sentence is drawing a comparison. *Than* shows a comparative relationship whereas *then* shows sequence or consequence. *A* and *C* can be eliminated because they contain the choice *then*. *B* is incorrect because *affect* is a verb while this sentence requires a noun.

11. A: The affix *circum–* originates from Latin and means *around or surrounding*. It is also related to other round words, such as circle and circus. The rest of the choices do not relate to the affix *circum–* and are therefore incorrect.

12. A: Only two of these suffixes, *–ize* and *–en*, can be used to form verbs, so *B* and *D* are incorrect. Those choices create adjectives. The suffix *–ize* means "to convert or turn into." The suffix *–en* means "to become." Because this sentence is about converting ideas into money, money + *–ize* or *monetize* is the most appropriate word to complete the sentence, so *C* is incorrect.

13. D: Students can use context clues to make a careful guess about the meaning of unfamiliar words. Although all of the words in a sentence can help contribute to the overall sentence, in this case, the adjective that pairs with *ubiquitous* gives the most important hint to the student—cars were first *rare*, but now they are *ubiquitous*. The inversion of *rare* is what gives meaning to the rest of the sentence and *ubiquitous* means "existing everywhere" or "not rare." *A* is incorrect because *ago* only indicates a time frame. *B* is incorrect because *cars* does not indicate a contrasting relationship to the word *ubiquitous* to provide a good context clue. *C* is incorrect because it also only indicates a time frame, but used together with *rare*, it provides the contrasting relationship needed to identify the meaning of the unknown word.

14. C: A style guide offers advice about proper formatting, punctuation, and usage when writing for a specific field, such as journalism or scientific research. The other resources would not offer similar information. A dictionary is useful for looking up definitions; a thesaurus is useful for looking up synonyms and antonyms. A grammar book is useful for looking up specific grammar topics. Thus, Choices *A*, *C*, and *D* are incorrect.

15. B: A word's origin is also known as its *etymology*. In addition to offering a detailed list of a word's various meanings, a dictionary also provides information about a word's history, such as when it first came into use, what language it originated from, and how its meaning may have changed over time. A thesaurus is for identifying synonyms and antonyms, so *A* is incorrect. A style guide provides formatting, punctuation, and syntactical advice for a specific field, and a grammar book is related to the appropriate placement of words and punctuation, which does not provide any insight into a word's meaning. Therefore, Choices *A*, *C*, and *D* are incorrect.

16. C: The .org domain on websites is generally used by nonprofit groups or community organizations. A government website uses .gov, and .edu is used for educational institutions. Private companies and businesses use .com, so Choices *A*, *B*, and *D* are incorrect.

17. B: The key word here is "inform," which is the primary purpose of all informative modes. They contain facts, definitions, instructions, and other elements with the objective purpose of informing a reader—such as study guides, instruction manuals, and textbooks. Choice *A* is incorrect because an argumentative mode contains language that is subjective and is intended to persuade or to inform with a persuasive bias. Choice *C* is incorrect as a narrative mode is used primarily to tell a story and has no intention of informing, nor is the language inherently objective. Choice *D* is incorrect as descriptive modes possess no inherent intent to inform, and are used primarily to describe.

18. A: Editorials, recommendation letters, and cover letters all seek to persuade a reader to agree with the author, which reflects an argumentative mode. Choice *B* is incorrect because the intent of the above examples is to persuade a reader to agree with the author, not to present information. Choice *C* is incorrect as the above examples are not trying to tell a story. Choice *D* is also incorrect because while the above examples may contain many descriptions, that is not their primary purpose.

19. D: Both the audience and primary purpose are important for choosing a writing mode. The audience is an important factor as the diction, tone, and stylistic choices of a written piece are tailored to fit the audience demographic. The primary purpose is the reason for writing the piece, so the mode of writing must be tailored to the most effective delivery method for the message. Choice *A* is incorrect because it only takes into account one of the aspects for choosing a mode and the audience, but leaves out the primary purpose. Choice *B* is incorrect for the same reason, except it only takes into account the primary purpose and forgets the audience. Choice *C* is incorrect as the main idea is the central theme or topic of the piece, which can be expressed in any form the author chooses. Because the mode depends on the reason the author wrote the piece, the main idea is not an important factor in determining which mode of writing to use.

20. C: Pathos is the rhetorical appeal that draws on an audience's emotions and sympathies. Choice *A* is incorrect as logos appeals to the audience's logic, reason, and rational thinking, using facts and definitions. Choice *B* is incorrect because ethos appeals to the audience's sense of ethics and moral obligations. Choice *D* is incorrect because *C* contains the correct answer; thus, the answer cannot be "None of the above."

21. C: The main idea of a piece is its central theme or subject and what the author wants readers to know or understand after they read. Choice *A* is incorrect because the primary purpose is the reason that a piece was written, and while the main idea is an important part of the primary purpose, the above elements are not developed with that intent. Choice *B* is incorrect because while the plot refers to the events that occur in a narrative, organization, tone, and supporting details are not used only to develop plot. Choice *D* is incorrect because characterization is the description of a person.

22. A: The passage describes a situation and then explains the causes that led to it. Also, it utilizes cause and effect signal words, such as *causes, factors, so,* and *as a result. B* is incorrect because a compare and contrast order considers the similarities and differences of two or more things. *C* is incorrect because spatial order describes where things are located in relation to each other. Finally, *D* is incorrect because time order describes when things occurred chronologically.

23. B: Children and less educated audiences tend to understand short, succinct sentences more effectively because their use helps increase information processing. Choice *A* is incorrect as longer, more fluid sentences are best used for adults and more educated audiences because they minimize processing times and allow for more information to be conveyed. Choices *C* and *D* are incorrect because there is no correlation between a given profession and a writing style; rather, it depends on how familiar the audience is with a given subject.

24. D: Relevant information refers to information that is closely related to the subject being researched. Students might get overwhelmed by information when they first begin researching, so they should learn how to narrow down search terms for their field of study. Both Choices *A* and *B* are incorrect because they start with a range that is far too wide; the student will spend too much time sifting through unrelated information to gather only a few related facts. Choice *C* introduces a more limited range, but it is not closely related to the topic that is being researched. Finally, Choice *D* is correct because the student is choosing books that are more closely related to the topic and is using the index or table of contents to evaluate whether the source contains the necessary information.

25. A: The publisher, author, and references are elements of a resource that determine credibility. If the publisher has published more than one work, the author has written more than one piece on the subject, or the work references other recognized research, the credibility of a source will be stronger. Choice *B* is incorrect because the subject and title may be used to determine relevancy, not credibility, and the audience does not have much to do with the credibility of a source. Choice *C* is incorrect because the organization, stylistic choices, and transition words are all components of an effectively-written piece, but they have less to do with credibility, other than to ensure that the author knows how to write. The length and tone of a piece are a matter of author's preference, and a work does not have to be written by multiple people to be considered a credible source.

26. B: In choice B the word *Uncle* should not be capitalized, because it is not functioning as a proper noun. If the word named a specific uncle, such as *Uncle Jerry*, then it would be considered a proper noun and should be capitalized. Choice *A* correctly capitalizes the proper noun *East Coast*, and does not capitalize *winter*, which functions as a common noun in the sentence. Choice *C* correctly capitalizes the name of a specific college course, which is considered a proper noun. Choice *D* correctly capitalizes the proper noun *Jersey Shore*.

27. C: In choice C, *avid* is functioning as an adjective that modifies the word photographer. *Avid* describes the photographer Julia Robinson's style. The words *time* and *photographer* are functioning as nouns, and the word *capture* is functioning as a verb in the sentence. Other words functioning as adjectives in the sentence include, *local*, *business*, and *spare*, as they all describe the nouns they precede.

28. A: Choice A is correctly punctuated because it uses a semicolon to join two independent clauses that are related in meaning. Each of these clauses could function as an independent sentence. Choice B is incorrect because the conjunction is not preceded by a comma. A comma and conjunction should be used together to join independent clauses. Choice C is incorrect because a comma should only be used to join independent sentences when it also includes a coordinating conjunction such as *and* or *so*. Choice D does not use punctuation to join the independent clauses, so it is considered a fused (same as a run-on) sentence.

29. C: Choice C is a compound sentence because it joins two independent clauses with a comma and the coordinating conjunction *and*. The sentences in Choices B and D include one independent clause and one dependent clause, so they are complex sentences, not compound sentences. The sentence in Choice A has both a compound subject, *Alex and Shane*, and a compound verb, *spent and took*, but the entire sentence itself is one independent clause.

30. A: Choice A uses incorrect subject-verb agreement because the indefinite pronoun *neither* is singular and must use the singular verb form *is*. The pronoun *both* is plural and uses the plural verb form of *are*. The pronoun *any* can be either singular or plural. In this example, it is used as a plural, so the plural verb form *are* is used. The pronoun *each* is singular and uses the singular verb form *is*.

31. B: Choice B is correct because the pronouns *he* and *I* are in the subjective case. *He* and *I* are the subjects of the verb *like* in the independent clause of the sentence. Choice A, C, and D are incorrect because they all contain at least one objective pronoun (*me* and *him*). Objective pronouns should not be used as the subject of the sentence, but rather, they should come as an object of a verb. To test for correct pronoun usage, try reading the pronouns as if they were the only pronoun in the sentence. For example, *he* and *me* may appear to be the correct answer choices, but try reading them as the only pronoun.

> He like[s] to go fishing…
>
> Me like to go fishing…
>
> When looked at that way, *me* is an obviously incorrect choice.

32. A: It is necessary to put a comma between the date and the year. It is also required to put a comma between the day of the week and the month. Choice B is incorrect because it is missing the comma between the day and year. Choice C is incorrect because it adds an unnecessary comma between the month and date. Choice D is missing the necessary comma between day of the week and the month.

33. D: *Discipline* and *perseverance* are both spelled correctly in choice *d*. These are both considered commonly misspelled words. One or both words are spelled incorrectly in choices A, B, and C.

34. D: In Choice D, the word function is a noun. While the word *function* can also act as a verb, in this particular sentence it is acting as a noun as the object of the preposition *at*. Choices A and B are incorrect because the word *function* cannot be used as an adjective or adverb.

35. C: Cacti is the correct plural form of the word *cactus*. Choice *A* (*tomatos*) includes an incorrect spelling of the plural of *tomato*. Both choice B (*analysis*) and choice D (*criterion*) are incorrect because they are in singular form. The correct plural form for these choices would be *criteria* and analyses.

36. B: Quotation marks are used to indicate something someone said. The example sentences feature a direct quotation that requires the use of double quotation marks. Also, the end punctuation, in this case a question mark, should always be contained within the quotation marks. Choice *A* is incorrect because there is an unnecessary period after the quotation mark. Choice *C* is incorrect because it uses single quotation marks, which are used for a quote within a quote. Choice *D* is incorrect because it places the punctuation outside of the quotation marks.

37. D: The prefix *trans* means across, beyond, over. Choices A, B, and C are incorrect because they are the meanings of other prefixes. Choice *A* is a meaning of the prefix *de*. Choice *B* is the meaning of the prefix *omni*. Choice *C* is one of the meanings of the prefix *pro*. The example words are helpful in determining the meaning of *trans*. All of the example words—*transfer, transact, translation, transport*—indicate something being *across, beyond,* or *over* something else. For example, *translation* refers to text going across languages. If no example words were given, you could think of words starting with *trans* and then compare their meanings to try to determine a common definition.

38. D: In choice D, the word *part* functions as an adjective that modifies the word *Irish*. Choices A and C are incorrect because the word *part* functions as a noun in these sentences. Choice *B* is incorrect because the word *part* functions as a verb.

39. C: *All of Shannon's family and friends* is the complete subject because it includes who or what is doing the action in the sentence as well as the modifiers that go with it. Choice *A* is incorrect because it only includes the simple subject of the sentence. Choices B and D are incorrect because they only include part of the complete subject.

40. D: Choice *D* directly addresses the reader, so it is in second person point of view. This is an imperative sentence since it issues a command; imperative sentences have an *understood you* as the subject. Choice *A* uses first person pronouns *I* and *my*. Choices B and C are incorrect because they use third person point of view.

41. A: The word *prodigious* is defined as very impressive, amazing, or large. In this sentence, the meaning can be drawn from the words *they became nervous about climbing its distant peak*, as this would be an appropriate reaction upon seeing a very large peak that's far in the distance. Choices B, C, and D do not accurately define the word *prodigious*, so they are incorrect.

42. D: Choice *D* correctly places a hyphen after the prefix *ex* to join it to the word *husband.* Words that begin with the prefixes *great, trans, ex, all,* and *self*, require a hyphen. Choices A and C place hyphens in words where they are not needed. *Beagle mix* would only require a hyphen if coming before the word *Henry*, since it would be serving as a compound adjective in that instance. Choice *B* contains hyphens that are in the correct place but are formatted incorrectly since they include spaces between the hyphens and the surrounding words.

43. C: Choice *C* is correct because quotation marks should be used for the title of a short work such as a poem. Choices A, B, and D are incorrect because the titles of novels, films, and newspapers should be placed in italics, not quotation marks.

44. C: This question focuses on the correct usage of the commonly confused word pairs of *it's/its* and *then/than*. *It's* is a contraction for *it is* or *it has*. *Its* is a possessive pronoun. The word *than* shows comparison between two things. *Then* is an adverb that conveys time. Choice *C* correctly uses *it's* and *than*. *It's* is a contraction for *it has* in this sentence, and *than* shows comparison between *work* and *rest*. None of the other answer choices use both of the correct words.

45. C: The suffix *-fy* means to make, cause, or cause to have. Choices A, B, and D are incorrect because they show meanings of other suffixes. Choice *A* shows the meaning of the suffix *-ous*. Choice *B* shows the meaning of the suffix *–ist*, and choice D shows the meaning of the suffix *-age*.

46. B: *Dissatisfied* and *accommodations* are both spelled correctly in Choice *B*. These are both considered commonly misspelled words. One or both words are spelled incorrectly in choices A, C, and D.

47. B: Choice *B* is an imperative sentence because it issues a command. In addition, it ends with a period, and an imperative sentence must end in a period or exclamation mark. Choice *A* is a declarative sentence that states a fact and ends with a period. Choice *C* is an exclamatory sentence that shows strong emotion and ends with an exclamation point. Choice *D* is an interrogative sentence that asks a question and ends with a question mark.

48. A: *After* and *then* are transitional words that indicate time or position. Choice *B* is incorrect because the words *at, with,* and *to* are used as prepositions in this sentence, not transitions. Choice *C* is incorrect because the words *had* and *took* are used as verbs in this sentence. In Choice *D*, *a* and *the* are used as articles in the sentence.

49. C: Choice *C* is a compound sentence because it joins two independent clauses—*The baby was sick* and *I decided to stay home from work*—with a comma and the coordinating conjunction *so*. Choices A, B, and D, are all simple sentences, each containing one independent clause with a complete subject and predicate. Choices A and D each contain a compound subject, or more than one subject, but they are still simple sentences that only contain one independent clause. Choice *B* contains a compound verb (more than one verb), but it's still a simple sentence.

50. A: Choice *A* includes all of the words functioning as nouns in the sentence. Choice *B* is incorrect because it includes the words *final* and *academic,* which are functioning as adjectives in this sentence. The word *he* makes Choice *C* incorrect because it is a pronoun. This example also includes the word *library*, which can function as a noun, but is functioning as an adjective modifying the word *databases* in this sentence. Choice *D* is incorrect because it leaves out the proper noun *Robert*.

51. C: The simple subject of this sentence, the word *lots*, is plural. It agrees with the plural verb form *were*. Choice *A* is incorrect, because the simple subject *there*, referring to the two constellations, is considered plural. It does not agree with the singular verb form *is*. In Choice *B*, the singular subject *four*, does not agree with the plural verb form *needs*. In Choice *D* the plural subject *everyone* does not agree with the singular verb form *have*.

52. B: *Excellent* and *walking* are adjectives modifying the noun *tours*. *Rich* is an adjective modifying the noun *history*, and *brotherly* is an adjective modifying the noun *love*. Choice *A* is incorrect because all of these words are functioning as nouns in the sentence. Choice *C* is incorrect because all of these words are functioning as verbs in the sentence. Choice *D* is incorrect because all of these words are considered prepositions, not adjectives.

53. D: The object pronouns *her* and *me* act as the indirect objects of the sentence. If *me* is in a series of object pronouns, it should always come last in the series. Choice *A* is incorrect because it uses subject pronouns *she* and *I*. Choice *B* is incorrect because it uses the subject pronoun *she*. Choice *C* uses the correct object pronouns, but they are in the wrong order.

54. A: In this example, a colon is correctly used to introduce a series of items. Choice *B* places an unnecessary comma before the word *because*. A comma is not needed before the word *because* when it introduces a dependent clause at the end of a sentence and provides necessary information to understand the sentence. Choice *C* is incorrect because it uses a semi-colon instead of a comma to join a dependent clause and an independent clause. Choice *D* is incorrect because it uses a colon in place of a comma and coordinating conjunction to join two independent clauses.

55. B: Choice *B* correctly uses the contraction for *you are* as the subject of the sentence, and it correctly uses the possessive pronoun *your* to indicate ownership of the jacket. It also correctly uses the adverb *there*, indicating place. Choice *A* is incorrect because it reverses the possessive pronoun *your* and the contraction for *you are*. It also uses the possessive pronoun *their* instead of the adverb *there*. Choice *C* is incorrect because it reverses *your* and *you're* and uses the contraction for *they are* in place of the adverb *there*. Choice *D* incorrectly uses the possessive pronoun *their* instead of the adverb *there*.

Dear TABE Test Taker,

We would like to start by thanking you for purchasing this study guide for your TABE exam. We hope that we exceeded your expectations.

Our goal in creating this study guide was to cover all of the topics that you will see on the test. We also strove to make our practice questions as similar as possible to what you will encounter on test day. With that being said, if you found something that you feel was not up to your standards, please send us an email and let us know.

We would also like to let you know about other books in our catalog that may interest you.

Test Name	Amazon Link
GED	amazon.com/dp/1628459433
HiSET	amazon.com/dp/1628457694

We have study guides in a wide variety of fields. If the one you are looking for isn't listed above, then try searching for it on Amazon or send us an email.

Thanks Again and Happy Testing!
Product Development Team
info@studyguideteam.com

Interested in buying more than 10 copies of our product? Contact us about bulk discounts:

bulkorders@studyguideteam.com

FREE Test Taking Tips DVD Offer

To help us better serve you, we have developed a Test Taking Tips DVD that we would like to give you for FREE. **This DVD covers world-class test taking tips that you can use to be even more successful when you are taking your test.**

All that we ask is that you email us your feedback about your study guide. Please let us know what you thought about it – whether that is good, bad or indifferent.

To get your **FREE Test Taking Tips DVD**, email freedvd@studyguideteam.com with "FREE DVD" in the subject line and the following information in the body of the email:

     a. The title of your study guide.

     b. Your product rating on a scale of 1-5, with 5 being the highest rating.

     c. Your feedback about the study guide. What did you think of it?

     d. Your full name and shipping address to send your free DVD.

If you have any questions or concerns, please don't hesitate to contact us at freedvd@studyguideteam.com.

Thanks again!

CPSIA information can be obtained
at www.ICGtesting.com
Printed in the USA
LVHW060113061020
668039LV00018B/1482